Keynotes in Organic Chemistry

Andrew F. Parsons

Blackwell
Science

© 2003 by Blackwell Science Ltd,
a Blackwell Publishing Company
Editorial Offices:
9600 Garsington Road, Oxford OX4 2DQ,
UK
Tel: +44 (0) 1865 776868
Blackwell Publishing, Inc., 350 Main
Street, Malden, MA 02148-5018, USA
Tel: +1 781 388 8250
Iowa State Press, a Blackwell Publishing
Company, 2121 State Avenue, Ames,
Iowa 50014-8300, USA
Tel: +1 515 292 0140
Blackwell Publishing Asia Pty Ltd,
550 Swanston Street, Carlton South,
Victoria 3053, Australia
Tel: +61 (0)3 9347 0300
Blackwell Wissenschafts Verlag,
Kurfürstendamm 57, 10707 Berlin,
Germany
Tel: +49 (0)30 32 79 060

First published 2003 by Blackwell
Science Ltd

Library of Congress
Cataloging-in-Publication Data
is available

ISBN 0-632-05816-1

A catalogue record for this title
is available from the British Library

Set in 9/10 Times
by Integra Software Services Pvt. Ltd,
Pondicherry, India
Printed and bound in Great Britain by
MPG Books Ltd, Bodmin, Cornwall

For further information on
Blackwell Publishing, visit our website:
www.blackwellpublishing.com

CONTENTS

PREFACE

With the advent of modularisation and an ever-increasing number of examinations, there is a growing need for concise revision notes that encapsulate the key points of a subject in a meaningful fashion. This keynote revision guide provides concise organic chemistry notes for first year students studying chemistry and related courses (including biochemistry) in the UK. The text will also be appropriate for students on similar courses in other countries.

An emphasis is placed on presenting the material pictorially (pictures speak louder than words), and hence, there are relatively few paragraphs of text but many diagrams. These are annotated with key phrases that summarise important concepts/key information, and bullet points are included to concisely highlight key principles and definitions.

The material is organised to provide a structured programme of revision. Fundamental concepts, such as structure and bonding, functional group identification and stereochemistry, are introduced in the first three chapters. An important chapter on reactivity and mechanism is included to provide a short overview of the basic principles of organic reactions. The aim here is to provide the reader with a summary of the 'key tools' which are necessary for understanding the following chapters, and an emphasis is placed on the organisation of material based on reaction mechanism. Thus, an overview of general reaction pathways/ mechanisms (such as substitution and addition) is included, and these mechanisms are revisited in more detail in the following chapters. Chapters 5–10 are treated as essentially 'case studies', reviewing the chemistry of the most important functional groups. Alkyl halides are discussed first, and as these compounds undergo elimination reactions this is followed by the (electrophilic addition) reactions of alkenes. This leads on to the contrasting (electrophilic substitution) reactivity of benzene and derivatives in Chapter 7, while the rich chemistry of carbonyl compounds is divided into Chapters 8 and 9. This division is made on the basis of the different reactivity (addition versus substitution) of aldehydes/ketones and carboxylic acid derivatives to nucleophiles. A chapter is included to revise the importance of spectroscopy in structure elucidation, and finally, the structure and reactivity of a number of important natural products and synthetic polymers is highlighted in Chapter 11. Questions are included at the end of each chapter to test the reader's understanding, and outline answers are provided for all of the questions. Tables of useful physical data are included in the appendices at the back of the book.

There are numerous people whom I would like to thank for their help with this project. They include many students and colleagues at York, including Richard Taylor, Peter O'Brien, Bruce Gilbert, John Holman and especially David Smith. Their constructive comments were invaluable. I would also like to thank Sue and Thomas for their support and patience throughout this project. Finally, I would like to thank Anne Stanford for getting the project started and Paul Sayer, from Blackwell Publishing, for all his help in moving the project forward.

Dr Andrew F. Parsons
2002

1. STRUCTURE AND BONDING

Key point. Organic chemistry is the study of carbon compounds. *Ionic* bonds involve elements gaining or losing electrons, but the carbon atom is able to form four *covalent* bonds by sharing the four electrons in its outer shell. Single, double or triple bonds to carbon are possible. When carbon is bonded to a different element, the electrons are not shared equally as *electronegative* atoms (or groups) attract the electron density, whereas *electropositive* atoms (or groups) repel the electron density. An understanding of the electron-withdrawing or -donating ability of atoms, or a group of atoms, can be used to predict if an organic compound is a good *acid* or *base*. It is also helpful in understanding organic reactivity (see Chapter 4).

1.1 Ionic, covalent, coordinate and hydrogen bonds

- *Ionic bonds* are formed between molecules or between atoms with opposite charges. The negatively charged anion will electrostatically attract the positively charged cation. This is present in salts.

$$\text{Cation} \overset{\oplus}{\,}\,\text{\tiny{..............}}\,\overset{\ominus}{}\text{Anion} \qquad \text{e.g.} \qquad \text{Na} \overset{\oplus}{\,}\,\text{\tiny{..............}}\,\overset{\ominus}{\text{Cl}}$$

- *Covalent bonds* are formed when a pair of electrons is shared between two atoms. *Each* atom donates one electron, and a single line represents the two-electron bond.

$$\text{Atom}\!-\!\!\text{Atom} \qquad \text{e.g.} \qquad \text{Cl}\!-\!\text{Cl} \;\equiv\; \overset{\circ\circ}{\underset{\circ\circ}{\overset{\circ}{\underset{\circ}{\text{Cl}}}}} \; \overset{\circ\circ}{\underset{\circ\circ}{\overset{\circ}{\underset{\circ}{\text{Cl}}}}} \, {}^{\circ}_{\circ}$$

- *Coordinate* (or *dative*) *bonds* are formed when a pair of electrons is shared between two atoms. *One* atom donates both electrons, and a single line or an arrow represents the two-electron bond.

electron acceptor

O⟶ BH₃ or *electron donor*

$$\overset{\oplus}{\text{O}}\!-\!\overset{\ominus}{\text{BH}_3}$$

- *Hydrogen bonds* are formed when the partially positive (δ+) hydrogen of one molecule interacts with the partially negative (δ−) heteroatom (e.g. oxygen or nitrogen) of another molecule.

$$\text{Molecule–H}^{\delta+} \cdots\cdots \text{Heteroatom–Molecule} \quad \text{e.g.} \quad \text{HO}-\text{H}^{\delta+} \cdots\cdots \text{OH}_2$$

1.2   The octet rule

To form organic compounds, the carbon atom shares electrons to give a stable 'full shell' electron configuration of eight valence electrons.

Methane (CH₄)

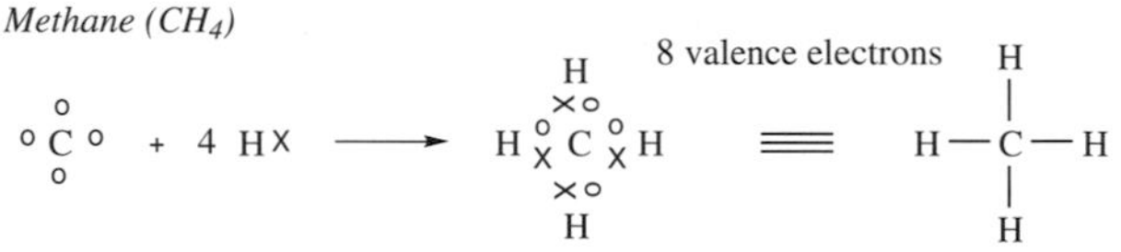

Lewis structure   Kekulé structure
(a line = 2 electrons)

C is in group 14 and hence has 4 valence electrons
H is in group 1 and hence has 1 valence electron

A single bond contains two electrons, a double bond contains four electrons and a triple bond contains six electrons. A lone (or non-bonding) pair of electrons is represented by two dots.

Carbon dioxide (CO₂)   *Hydrogen cyanide (HCN)*

$$:\!O\!=\!C\!=\!O\!:$$ $$H-C\equiv N:$$

1.3   Formal charge

Formal positive or negative charges are assigned to atoms, which have an apparent 'abnormal' number of bonds.

Atom(s)	C	N, P	O, S	F, Cl, Br, I
Group number	14	15	16	17
Normal number of 2 electron bonds	4	3	2	1

Formal charge =	group number in periodic table	−	number of bonds to atom	−	number of unshared electrons	−	10

Example: nitric acid (HNO₃)

$$
\begin{array}{c}
\vdots\text{O}\text{:}\\
\vdots\text{O}-\text{N}{\overset{\oplus}{}}\nearrow\\
\text{H}\diagup\quad\text{:O:}^{\ominus}
\end{array}
$$

Nitrogen with 4 covalent bonds has a formal charge of +1

Formal charge: $15 - 4 - 0 - 10 = +1$

The nitrogen atom donates a pair of electrons to make this bond

Carbon forms four covalent bonds. When only three covalent bonds are present, the carbon atom can have either a formal negative charge or a formal positive charge.

- *Carbanions* – three covalent bonds to carbon and a formal negative charge.

Formal charge on C:

$14 - 3 - 2 - 10 = -1$

$$
\begin{array}{c}
\text{R}\diagdown\;{\overset{\ominus}{\underset{\cdot\cdot}{\text{C}}}}\diagup\text{R}\\
|\\
\text{R}
\end{array}
$$

8 outer electrons:
3 two-electron bonds
and 2 non-bonding
electrons

The negative charge is used to show the 2 non-bonding electrons

- *Carbocations* – three covalent bonds to carbon and a formal positive charge.

Formal charge on C:

$14 - 3 - 0 - 10 = +1$

$$
\begin{array}{c}
\text{R}\diagdown\;{\overset{\oplus}{\text{C}}}\diagup\text{R}\\
|\\
\text{R}
\end{array}
$$

6 outer electrons:
3 two-electron bonds

The positive charge is used to show the absence of 2 electrons

1.4 Sigma (σ-) and pi (π-) bonds

The electrons shared in a covalent bond result from an overlap of atomic orbitals to give a new molecular orbital. Electrons in the 1s- and 2s-orbitals combine to give sigma (σ-) bonds.

When two 1s-orbitals combine *in-phase*, this produces a *bonding molecular orbital*.

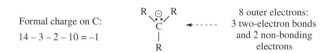

| s-orbital | s-orbital | bonding molecular orbital |

When two 1s-orbitals combine *out-of-phase*, this produces an *anti-bonding molecular orbital*.

Electrons in the p-orbitals can combine to give σ- or π-bonds.

- *Sigma (σ-) bonds* are strong bonds formed by head-on overlap of two atomic orbitals.

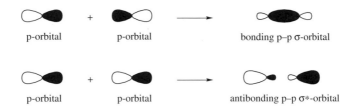

- *Pi (π-) bonds* are weaker bonds formed by side-on overlap of two p-orbitals.

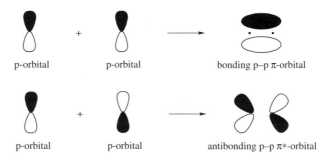

Only σ- or π-bonds are present in organic compounds. All single bonds are σ-bonds, while all multiple (double or triple) bonds are composed of one σ-bond and one or two π-bonds.

1.5 Hybridisation

- The ground-state electronic configuration of carbon is $1s^2 2s^2 2p_x^1 2p_y^1$.
- The six electrons fill up lower energy orbitals before entering higher energy orbitals (Aufbau principle).
- Each orbital is allowed a maximum of two electrons (Pauli exclusion principle).
- The two 2p electrons occupy separate orbitals before pairing up (Hund's rule).

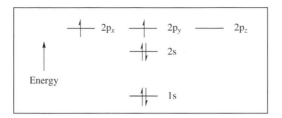

The carbon atom can mix the 2s and 2p atomic orbitals to form new hybrid orbitals in a process known as *hybridisation*.

- *sp³ Hybridisation.* For four single σ-bonds – carbon is sp³ hybridised (e.g. in methane, CH_4). The orbitals move as far apart as possible, and the lobes point to the corners of a tetrahedron (109.5° bond angle).

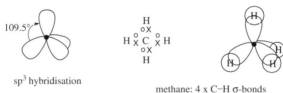

sp³ hybridisation

methane: 4 x C–H σ-bonds

- *sp² Hybridisation.* For three single σ-bonds and one π-bond – the π-bond requires one p-orbital, and hence the carbon is sp² hybridised (e.g. in ethene, $H_2C{=}CH_2$). The three sp²-orbitals point to the corners of a triangle (120° bond angle), and the remaining p-orbital is perpendicular to the sp² plane.

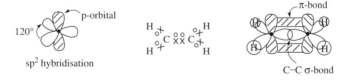

ethene: 4 x C–H σ-bonds, 1 x C–C σ-bond, 1 x C–C π-bond

- *sp Hybridisation.* For two single σ-bonds and two π-bonds – the two π-bonds require two p-orbitals, and hence the carbon is sp hybridised (e.g. in ethyne, HC≡CH). The two sp-orbitals point in the opposite directions (180° bond angle), and the two p-orbitals are perpendicular to the sp plane.

ethyne: 2 x C–H σ-bonds, 1 x C–C σ-bond, 2 x C–C π-bonds

- For a single C–C or C–O bond, the atoms are sp^3 hybridised and the carbon atom(s) is tetrahedral.
- For a double C=C or C=O bond, the atoms are sp^2 hybridised and the carbon atom(s) is trigonal planar.
- For a triple C≡C or C≡N bond, the atoms are sp hybridised and the carbon atom(s) is linear.

$3 = sp^3 \quad 2 = sp^2 \quad 1 = sp$

The shape of organic molecules is therefore determined by the hybridisation of the atoms.

Functional groups (see Section 2.1) which contain π-bonds are generally more reactive as the π-bond is weaker than the σ-bond. The π-bond in an alkene or alkyne is around 210–230 kJ mol^{-1}, while the σ-bond is around 350 kJ mol^{-1}.

Bond	Bond strength (kJ mol^{-1})	Bond length (nm)
C—C	350	0.15
C=C	560	0.13
C≡C	810	0.12

The shorter the bond length, the stronger the bond. For C–H bonds, the greater the 's' character of the carbon orbitals, the shorter the bond length, because the electrons are held closer to the nucleus.

$$\underset{\text{longest}}{\overset{sp^3}{H_3C-CH_2 \top H}} \quad > \quad \overset{sp^2}{H_2C=CH-H} \quad > \quad \underset{\text{shortest}}{\overset{sp}{HC\equiv C\top H}}$$

A single C–C σ-bond can undergo free rotation at room temperature, but a π-bond prevents free rotation around a C=C bond. For the maximum orbital overlap in a π-bond, the two p-orbitals need to be parallel to one another. Any rotation around the C=C bond will break the π-bond.

1.6 Inductive effects, hyperconjugation and mesomeric effects

1.6.1 Inductive effects

In a covalent bond between two different atoms, the electrons in the σ-bond are not shared equally. The electrons are attracted towards the most electronegative atom. An arrow drawn above the line representing the covalent bond can show this. (Sometimes an arrow is drawn on the line.) Electrons are pulled in the direction of the arrow.

When the atom (X) is more electronegative than carbon	*When the atom (Z) is less electronegative than carbon*
electrons **attracted to X**	electrons **attracted to C**
$\underset{\displaystyle \overset{\|}{}}{\overset{\displaystyle \overrightarrow{+}}{\underset{}{\searrow\atop\nearrow}\underset{\delta+}{C}-\underset{\delta-}{X}}}$	$\underset{\displaystyle \overset{\|}{}}{\overset{\displaystyle \overleftarrow{+}}{\underset{}{\searrow\atop\nearrow}\underset{\delta-}{C}-\underset{\delta+}{Z}}}$
negative inductive effect. –I	positive inductive effect. +I

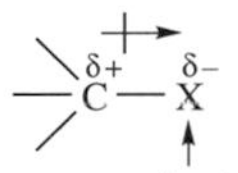

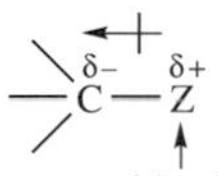

–I groups	+I groups
X = Br, Cl, NO$_2$, OH, OR, SH, SR, NH$_2$, NHR, NR$_2$, CN, CO$_2$H, CHO, C(O)R	Z = R (alkyl or aryl), metals (e.g. Li or Mg)
The more electronegative the atom (X), the stronger the –I effect	The more electropositive the atom (Z), the stronger the +I effect

Pauling electronegativity scale	The inductive effect of the atom rapidly diminishes as the chain length increases
K = 0.8 I = 2.5 C = 2.5 Br = 2.8 N = 3.0 Cl = 3.0 O = 3.5 F = 4.0	$\underset{\delta\delta\delta+}{H_3C}-\underset{\delta\delta+}{CH_2}-\underset{\delta+}{CH_2}-\overset{\longrightarrow}{\underset{\delta-}{CH_2}}-Cl$
The higher the value, the more electronegative the atom	experiences a negligible –I effect experiences a strong –I effect

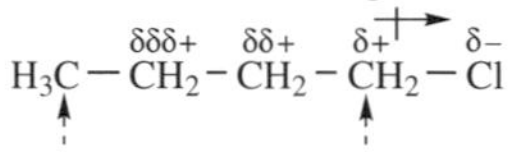

The overall polarity of a molecule is determined by the individual bond polarities, formal charges and lone pair contributions, and this can be measured by the dipole moment (μ). The higher the dipole moment (measured in debyes (D)), the more polar the compound.

1.6.2 Hyperconjugation

A σ-bond can stabilise a neighbouring carbocation (or positively charged carbon) by donating electrons to the vacant p-orbital. The positive charge is delocalised or 'spread out', and this stabilising effect is known as *resonance*.

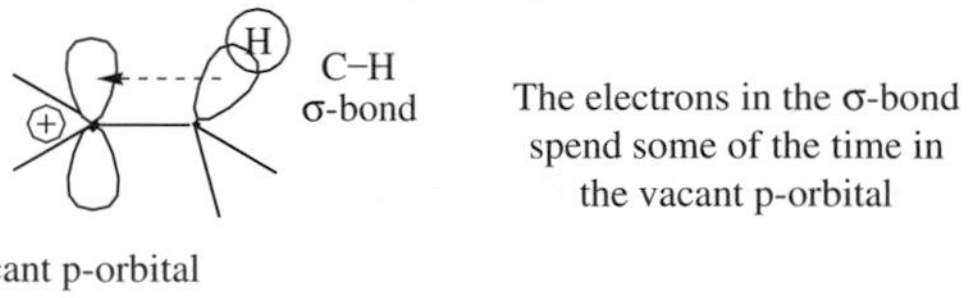

C–H σ-bond

The electrons in the σ-bond spend some of the time in the vacant p-orbital

vacant p-orbital

1.6.3   Mesomeric effects

Whilst inductive effects pull electrons through the σ-bond framework, electrons can also move through the π-bond network. A π-bond can stabilise a negative charge, a positive charge, a lone pair of electrons or an adjacent bond by *resonance* (i.e. delocalisation or 'spreading out' of the electrons). Curly arrows (see Section 4.1) are used to represent the movement of π- or non-bonding electrons to give different resonance forms. It is only the electrons, not the nuclei, that move in the resonance forms, and a double-headed arrow is used to show their relationship.

Positive mesomeric effect

- When a π-system donates electrons, the π-system has a positive mesomeric effect, +M.

donates electrons:
+M group

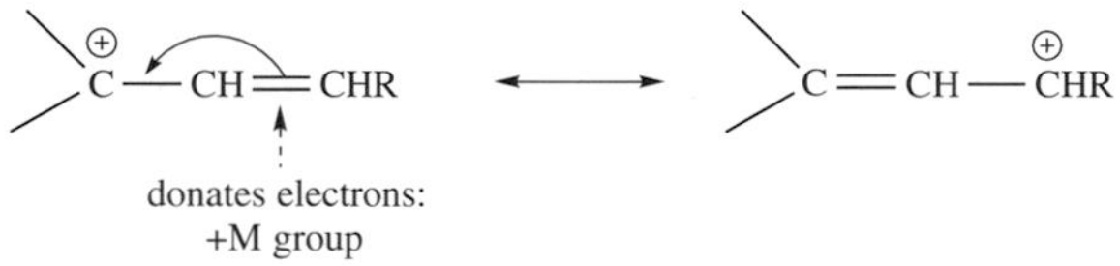

- When a lone pair of electrons is donated, the group donating the electrons has a positive mesomeric effect, +M.

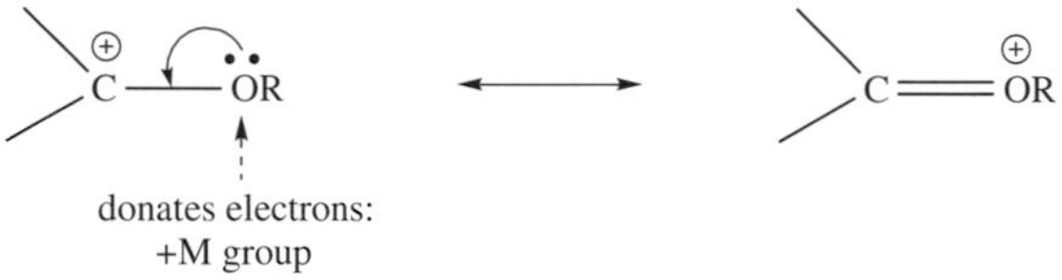

donates electrons:
+M group

Negative mesomeric effect

- When a π-system accepts electrons, the π-system has a negative mesomeric effect, −M.

accepts electrons:
−M group

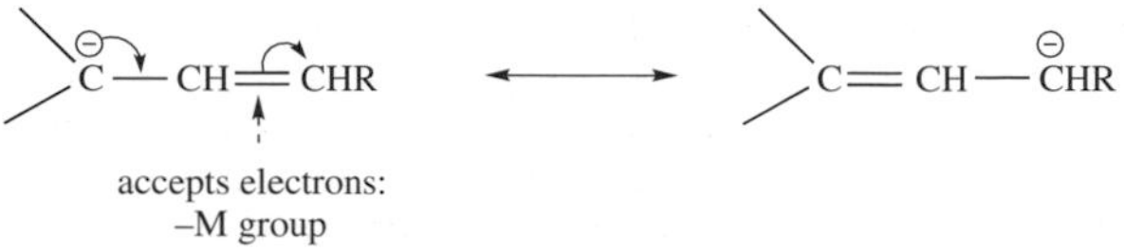

The actual structures of the cations or anions lie somewhere between the two resonance forms. All resonance forms must have the same overall charge and obey the same rules of valency.

−M groups generally contain an electronegative atom(s) and/or a π-bond(s):

CHO, C(O)R, CO_2H, CO_2Me, NO_2, CN, aromatics, alkenes

+M groups generally contain a lone pair of electrons or a π-bond(s):

$\ddot{C}l$, $\ddot{B}r$, $\ddot{O}H$, $\ddot{O}R$, $\ddot{S}H$, $\ddot{S}R$, $\ddot{N}H_2$, $\ddot{N}HR$, $\ddot{N}R_2$, aromatics, alkenes

Aromatic (or aryl) groups and alkenes can be *both* +M and −M

In neutral compounds, there will always be a +M *and* −M group(s): one group donates (+M) the electrons and the other group(s) accepts the electrons (−M).

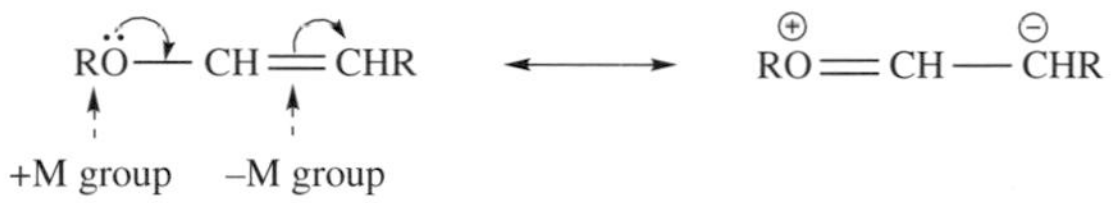

All resonance forms are *not* of the same energy. In phenol, for example, the resonance form with the intact aromatic benzene ring (see Section 7.1) is expected to predominate.

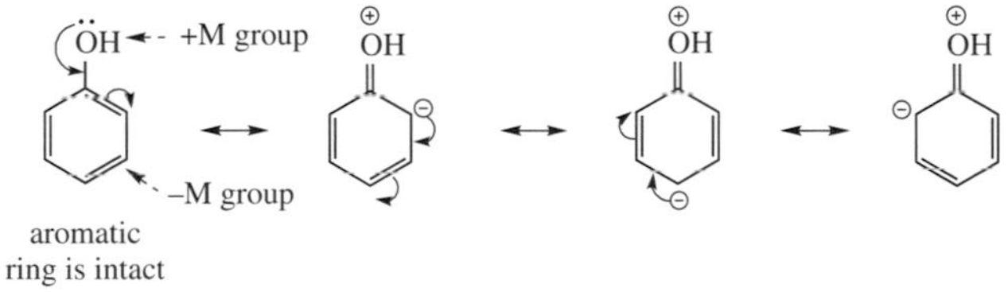

As a rule of thumb, the more resonance structures an anion, cation or neutral π-system can have, the more stable it is.

Inductive versus mesomeric effects

Mesomeric effects are generally stronger than inductive effects. A +M group is likely to stabilise an anion more effectively than a +I group.

Mesomeric effects can be effective over much longer distances than inductive effects, provided that *conjugation* is present (i.e. alternating single and double bonds). Whereas inductive effects are determined by distance, mesomeric effects are determined by the relative positions of +M and −M groups in a molecule (see Section 1.7).

1.7   Acidity and basicity

1.7.1   Acids

An acid is a substance that donates a proton (Brønsted–Lowry). Acidic compounds have low pK_a-values and are good proton donors, as the anions (or conjugate bases), formed on deprotonation, are relatively stable.

In water:

acidity constant

$$HA \quad + \quad H_2\ddot{O} \quad \underset{}{\overset{K_a}{\rightleftharpoons}} \quad H_3O^{\oplus} \quad + \quad A^{\ominus}$$

Acid   Base   conjugate   conjugate
                                            acid   base

The more stable the conjugate base the stronger the acid

$$K_a \approx \frac{[H_3O^{\oplus}][A^{\ominus}]}{[HA]}$$

As H_2O is in excess

$pK_a = -\log_{10}K_a$

The higher the value of K_a, the lower the pK_a-value and the more acidic is HA

The pK_a-value equals the pH of the acid when it is half dissociated. At pHs above the pK_a the acid exists predominantly as the conjugate base in water. At pHs below the pK_a, it exists predominantly as HA.

pH = 0, strongly acidic

pH = 7, neutral

pH = 14, strongly basic

The pK_a-values are influenced by the solvent. Polar solvents will stabilise cations and/or anions by *solvation*, in which the charge is delocalised over the solvent (e.g. by hydrogen-bonding in water).

$$HO-H^{\delta+} \cdots A^{\ominus} \cdots {}^{\delta+}H-OH$$

$$H_2O^{\delta-} \cdots \overset{\overset{\displaystyle H}{|}}{O^{\oplus}} \cdots {}^{\delta-}OH_2$$
$$\underset{H \quad H}{}$$

The more electronegative the atom bearing the negative charge, the more stable the conjugate base (which is negatively charged).

pK_a	3		16		33		48
most acidic	HF	>	H_2O	>	NH_3	>	CH_4

increasing electronegativity

Therefore, F^- is more stable than H_3C^-.

The conjugate base can also be stabilised by −I and −M groups which can delocalise the negative charge. (The more 'spread out' the negative charge, the more stable it is.)

> −I and −M groups therefore *lower* the pK_a, while
> +I and +M groups *raise* the pK_a

Inductive effects and carboxylic acids

The carboxylate anion is formed on deprotonation of carboxylic acids. The anion is stabilised by resonance (i.e. the charge is spread over both oxygen atoms) but can also be stabilised by the R group if this has a −I effect.

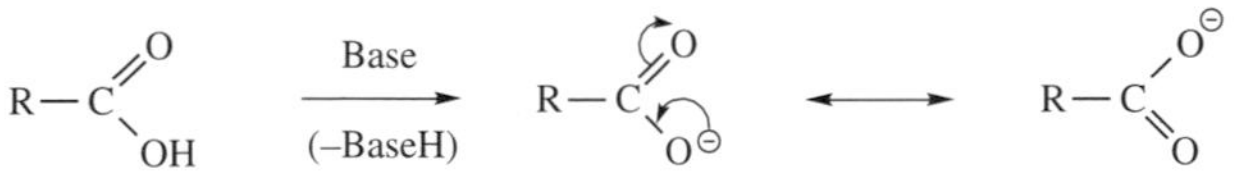

The greater the −I effect, the more stable the carboxylate anion and the more acidic the carboxylic acid.

$$F-CH_2-CO_2H \qquad Br-CH_2-CO_2H \qquad H_3C-CO_2H$$

pK_a 2.7   2.9   4.8

Most acidic as F is more electronegative than Br and hence has a greater −I effect

Least acidic as the CH_3 group is a +I group

Inductive and mesomeric effects and phenols

Mesomeric effects can also stabilise positive and negative charges.

> The *negative* charge needs to be on an adjacent carbon atom for a −**M** group to stabilise it
>
> The *positive* charge needs to be on an adjacent carbon atom for a +**M** group to stabilise it

On deprotonation of phenol the phenoxide anion is formed. This is stabilised by the delocalisation of the negative charge onto the 2-, 4- and 6-positions of the benzene ring.

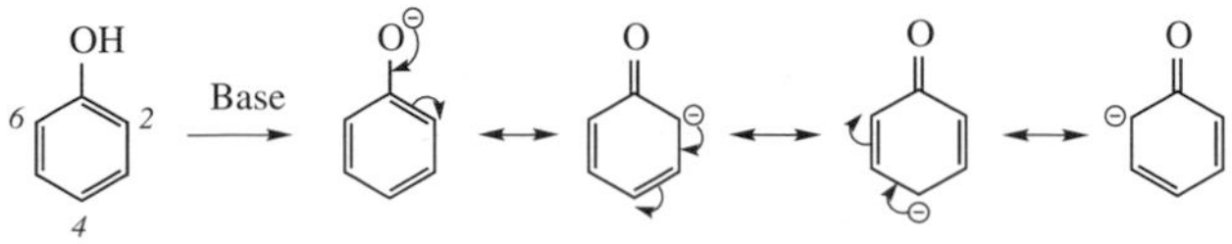

- If −M groups are introduced at the 2-, 4- and/or 6-positions, the anion can be further stabilised by delocalisation through the π-system, as the negative charge can be spread onto the −M group. We can use double-headed curly arrows to show this process (see Section 4.3).
- If −M groups are introduced at the 3- and/or 5-positions, the anion cannot be stabilised by delocalisation, as the negative charge cannot be spread onto the −M group. There is no way of using curly arrows to delocalise the charge onto the −M group.
- If −I groups are introduced on the benzene ring, the effect will depend on their distance from the negative charge. The closer the −I group is to the negative charge, the greater the stabilising effect will be. The order of −I stabilisation is therefore 2-position > 3-position > 4-position.
- The −M effects are much stronger than −I effects (see Section 1.6.3).

Examples

The NO₂ group is strongly electron-withdrawing; −I and −M

OH	OH	OH	OH, NO₂
pKₐ 9.9	8.4	7.2	4.0
Least acidic as no −I or −M groups on the ring	The NO₂ can only stabilise the anion inductively	The NO₂ can stabilise the anion inductively and by resonance	Most acidic as both NO₂ groups can stabilise the anion inductively and by resonance

1.7.2 Bases

A base is a substance that accepts a proton (Brønsted–Lowry). Basic compounds have high pK_a-values and are good proton acceptors, as the cations (or conjugate acids), formed on protonation, are relatively stable.

In water

$$B \; + \; H_2O \; \underset{}{\overset{K_b}{\rightleftharpoons}} \; \overset{\oplus}{B}H \; + \; H\overset{\ominus}{O}$$

basicity constant

| base | acid | | conjugate acid | conjugate base |

The strength of bases is usually described by the K_a- and pK_a-values of the conjugate acid.

$$\overset{\oplus}{BH} \quad + \quad H_2O \quad \underset{}{\overset{K_a}{\rightleftharpoons}} \quad \overset{..}{B} \quad + \quad H_3\overset{\oplus}{O}$$

$$K_a \approx \frac{[B]\,[H_3\overset{\oplus}{O}]}{[\overset{\oplus}{BH}]}$$

As H_2O is in excess

- If B is a *strong* base, then BH^+ will be relatively stable and not easily deprotonated. BH^+ will therefore have a *high* pK_a-value.
- If B is a *weak* base, then BH^+ will be relatively unstable and easily deprotonated. BH^+ will therefore have a *low* pK_a-value.

The cation can be stabilised by +I and +M groups, which can delocalise the positive charge. (The more 'spread out' the positive charge, the more stable it is.)

Inductive effects and aliphatic (or alkyl) amines

On protonation of amines, ammonium salts are formed.

$$R \overset{..}{-} NH_2 \quad + \quad H - X \quad \rightleftharpoons \quad R - \overset{\oplus}{NH_3} \quad X^{\ominus}$$

The greater the +I effect of the R group, the greater the electron density at nitrogen and the more basic the amine. The greater the +I effect, the more stable the ammonium cation and the more basic the amine.

pK_a 9.3	10.7	10.9	10.9

The pK_a-values *should* increase steadily as more +I alkyl groups are introduced on nitrogen. However, the pK_a-values are determined in *water*, and the more hydrogen atoms on the positively charged nitrogen, the greater the extent of hydrogen-bonding between water and the cation. This solvation leads to the stabilisation of the cations containing N–H bonds.

In organic solvents (which cannot solvate the cation), the order of pK_as is expected to be as follows.

$$R_3N \; > \; R_2NH \; > \; RNH_2 \; > \; NH_3 \qquad (R = +I \text{ alkyl group})$$

most basic   least basic

The presence of −I and/or −M groups on nitrogen reduces the basicity, and hence, for example, amides are poor bases.

Ethanamide has a pK_a of −0.5

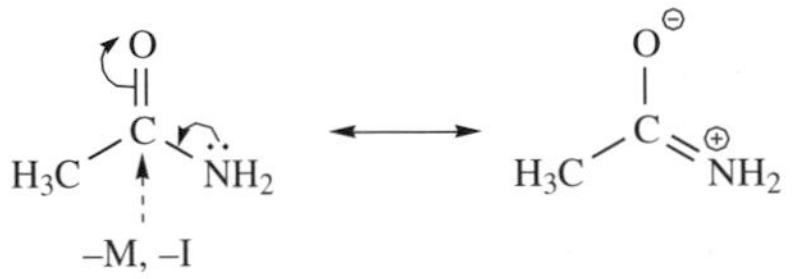

The C=O group stabilises the lone pair on nitrogen by resonance. This reduces the electron density on nitrogen

Mesomeric effects and aryl (or aromatic) amines

The lone pair of electrons on the nitrogen atom of aminobenzene (or aniline) can be stabilised by the delocalisation of the electrons onto the 2-, 4- and 6-positions of the benzene ring. Aromatic amines are therefore less basic than aliphatic amines.

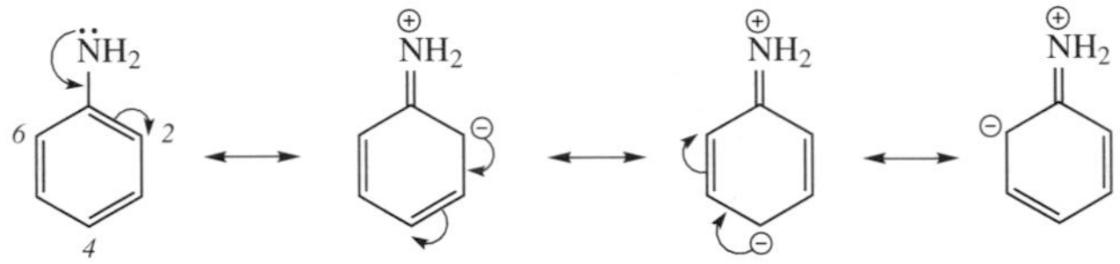

- If −M groups are introduced at the 2-, 4- and/or 6-positions (but not at the 3- or 5-position), the anion can be further stabilised by delocalisation, as the negative charge can be spread onto the −M group. This reduces the basicity of the amine.
- If −I groups are introduced on the benzene ring, the order of −I stabilisation is 2-position > 3-position > 4-position. This reduces the basicity of the amine.

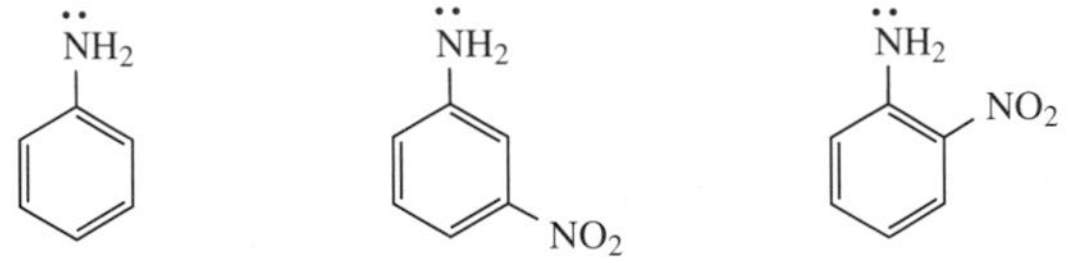

pK_a	4.6	2.45	−0.28
	Most basic as no −I or −M groups on the ring	The NO$_2$ can stabilise the lone pair inductively	Least basic. The NO$_2$ can stabilise the lone pair inductively and by resonance

- If +M groups (e.g. OMe) are introduced at the 2-, 4- or 6-position of aminobenzene, then the basicity is increased. This is because the +M group donates electron density to the carbon atom bearing the amine group.

The OMe group is –I but +M

$\ddot{N}H_2$ $\ddot{N}H_2$ $\ddot{N}H_2$

(three benzene ring structures with substituents: meta-OMe, ortho-OMe, para-:OMe)

pK$_a$ 4.2	4.5	5.3
Least basic. The OMe group cannot donate electron density to the carbon atom bearing the nitrogen	The OMe group can donate electron density to the nitrogen but it has a strong –I effect as it is in the 2-position	Most basic. The OMe group can donate electron density to the nitrogen and it has a weak –I effect (as well apart from the nitrogen)

Curly arrows can be used to show the delocalisation of electrons onto the carbon atom bearing the nitrogen.

$\ddot{N}H_2$ ⟷ $\ddot{N}H_2$ ⟷ $\ddot{N}H_2$ electron density adjacent to the nitrogen increases the basicity

(:OMe)   (⊕OMe)   (⊕OMe)

1.7.3 Lewis acids and bases

- A *Lewis acid* is any substance that accepts an electron pair in forming a coordinate bond (see Section 1.1). Examples include H$^+$, BF$_3$, AlCl$_3$, TiCl$_4$, ZnCl$_2$ and SnCl$_4$. They have unfilled valence shells and hence can accept electron pairs.
- A *Lewis base* is any substance that donates an electron pair in forming a coordinate bond. Examples include H$_2$O, ROH, RCHO, R$_2$CO, R$_3$N and R$_2$S. They all have a lone pair(s) of electrons on the heteroatom (O, N or S).

$$\begin{array}{c} R \\ \diagdown \\ \diagup \\ R \end{array} C = \ddot{O}\!: \qquad \begin{array}{c} Cl \\ | \\ Al - Cl \\ | \\ Cl \end{array} \quad \rightleftharpoons \quad \begin{array}{c} R \\ \diagdown \\ \diagup \\ R \end{array} C \overset{\oplus}{=} \ddot{O} - \overset{Cl}{\underset{Cl}{\overset{|}{\underset{|}{Al}}}} \overset{\ominus}{-} Cl$$

Lewis base   Lewis acid

1.7.4 Basicity and hybridisation

The greater the 's' character of an orbital, the lower in energy the electrons and the more tightly the electrons are held to the nucleus. The electrons in an sp-orbital are therefore less available for protonation than those in an sp^2- or sp^3-orbital, and hence the compounds are less basic.

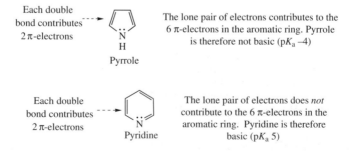

most basic $R_3\overset{\cdot\cdot}{N}$ $>$ $R_2C=\overset{\cdot\cdot}{N}H$ $>$ $RC\equiv\overset{\cdot\cdot}{N}$ least basic

most basic $R_3C^{\ominus}$ $>$ $R_2C=\overset{\ominus}{C}H$ $>$ $RC\equiv\overset{\ominus}{C}$ least basic

sp^3 sp^2 sp
(25% s)   (33% s)   (50% s)

1.7.5   Acidity and aromaticity

Aromatic compounds are planar, conjugated systems which have $4n+2$ electrons (Hückel's rule) (see Section 7.1). If, on deprotonation, the anion is part of an aromatic π-system, then the negative charge will be stabilised. Aromaticity will therefore *increase* the acidity of the compound.

fluorene
$pK_a = 22.8$

toluene
$pK_a = 40$

Each ring contributes
6 π-electrons

more stable

The anion can
be stabilised by
resonance and
it is aromatic
(planar and
14 π-electrons)

less stable

Although the anion
can be stabilised by
resonance it does not
contribute to the
aromaticity. (This
would give
8 π-electrons)

If a lone pair of electrons on a heteroatom is part of an aromatic π-system, then these electrons will not be available for protonation. Aromaticity will therefore *decrease* the basicity of the compound.

Each double
bond contributes
2 π-electrons

Pyrrole

The lone pair of electrons contributes to the
6 π-electrons in the aromatic ring. Pyrrole
is therefore not basic (pK_a −4)

Each double
bond contributes
2 π-electrons

Pyridine

The lone pair of electrons does *not*
contribute to the 6 π-electrons in the
aromatic ring. Pyridine is therefore
basic (pK_a 5)

1.7.6   Acid–base reactions

The pK_a-values can be used to predict if an acid–base reaction can take place. An acid will donate a proton to the conjugate base of any acid with a higher pK_a-value. This means that the product acid and base will be more stable than the starting acid and base.

$$HC\equiv C-H \quad + \quad {}^{\ominus}NH_2 \quad \rightleftharpoons \quad HC\equiv C^{\ominus} \quad + \quad NH_3$$

pK_a 25   Ammonia has a higher pK_a value than ethyne pK_a ~33
                 and hence the equilibrium lies to the right

Problems

(1) Using the I and M notations, identify the electronic effects of the following substituents.

     (a) −Me   (b) −Cl   (c) −NH$_2$ (d) −OH
     (e) −Br   (f) −CO$_2$Me   (g) −NO$_2$ (h) −CN

(2)   (a)   Use curly arrows to show how cations **A**, **B** and **C** (shown below) are stabilised by resonance, and draw the alternative resonance structure(s).
     (b)   Would you expect **A**, **B** or **C** to be the more stable? Briefly explain your reasoning.

    A   **B**   **C**

(3) Provide explanations for the following statements.
     (a)   The carbocation $CH_3OCH_2^+$ is more stable than $CH_3CH_2^+$.
     (b)   4-Nitrophenol is a much stronger acid than phenol (C_6H_5OH).
     (c)   The pK_a of CH_3COCH_3 is much lower than that of CH_3CH_3.
     (d)   The C–C single bond in CH_3CN is longer than that in $CH_2=CH-CN$.
     (e)   The cation $CH_2=CHCH_2^+$ is resonance stabilised, whereas the cation $CH_2=CHNMe_3^+$ is not.

(4) Why is cyclopentadiene (pK_a 15.5) a stronger acid than cycloheptatriene (pK_a ~36)?

(5) Which hydrogen atom would you expect to be the most acidic in each of the following compounds?
     (a)   4-Methylphenol (or p-cresol, 4-HOC$_6$H$_4$CH$_3$)
     (b)   4-Hydroxybenzoic acid (4-HOC$_6$H$_4$CO$_2$H)
     (c) $H_2C=CHCH_2CH_2C\equiv CH$
     (d) $HOCH_2CH_2CH_2C\equiv CH$

(6) Arrange the following sets of compounds in order of decreasing basicity. Briefly explain your reasoning.
     (a)   1-Aminopropane, ethanamide (CH_3CONH_2), guanidine [$HN=C(NH_2)_2$], aniline ($C_6H_5NH_2$).
     (b)   Aniline ($C_6H_5NH_2$), 4-nitroaniline, 4-methoxyaniline, 4-methylaniline.

2. FUNCTIONAL GROUPS, NOMENCLATURE AND DRAWING ORGANIC COMPOUNDS

Key point. Organic compounds are classified by *functional groups* which determine their chemistry. The names of organic compounds are derived from the functional group (or groups) and the main carbon chain. From the name, the structure of organic compounds can be drawn using *Kekulé, condensed* or *skeletal* structures.

2.1 Functional groups

A functional group is made up of an atom or atoms with characteristic chemical properties. The chemistry of organic compounds is determined by the functional groups that are present.

Hydrocarbons (only H and C present)

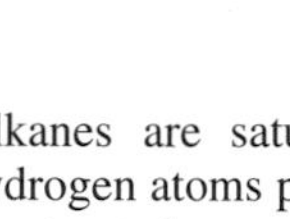

ethane	ethene	ethyne	benzene
(an alkane)	(an alkene)	(an alkyne)	(an arene)
single CC bond	double CC bond	triple CC bond	single/double CC bonds

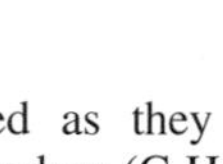

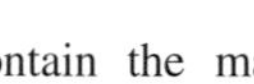

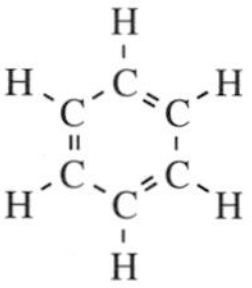

Alkanes are saturated as they contain the maximum number of hydrogen atoms per carbon (C_nH_{2n+2}). Alkenes, alkynes and arenes are unsaturated.

Carbon bonded to an electronegative atom(s)

- single bond (R = alkyl groups; see Section 2.2)

R—X	R—OH	R—O—R
Halide	Alcohol	Ether
X = Br, Cl, I		
R—NO$_2$	R—SH	R—S—R
Nitro	Thiol	Sulfide (thioether)

Amines

R—NH$_2$	primary amine
R—NHR	secondary amine
R—NR$_2$	tertiary amine
R—$\overset{\oplus}{N}R_3$	quaternary ammonium ion

- double bond to oxygen (these are known as carbonyl compounds)

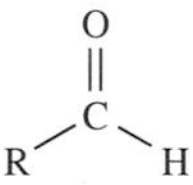

| Aldehyde | Ketone | Acid halide
X = Br, Cl | Carboxylic acid |

| Ester | Acid anhydride | Amide |

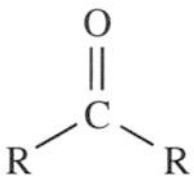

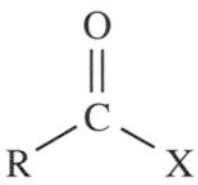

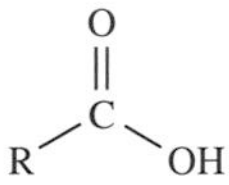

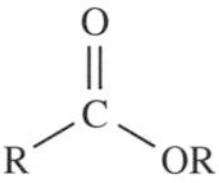

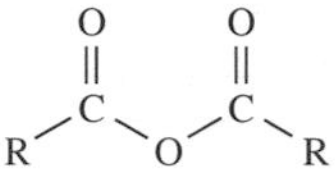

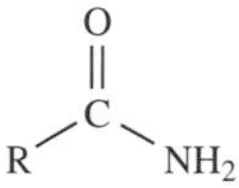

- triple bond to nitrogen

$$R - C \equiv N$$
Nitrile

2.2   Alkyl and aryl groups

When a hydrogen atom is removed from an alkane, this is an alkyl group. The symbol R is used to represent a general alkyl group (i.e. a methyl, ethyl, propyl, etc. group).

name (symbol)	structure	name (symbol)	structure
methyl (Me)	— CH_3	propyl (Pr)	— $CH_2CH_2CH_3$
ethyl (Et)	— CH_2CH_3	butyl (Bu)	— $CH_2CH_2CH_2CH_3$

When a hydrogen atom is removed from a benzene ring, this is a phenyl group. The symbol Ph is used to represent this.

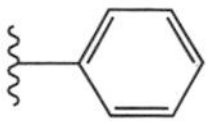

phenyl (C_6H_5), Ph

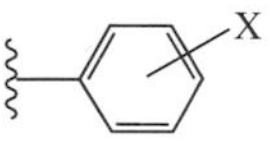

aryl, Ar
X = various
functional group(s)

Do not confuse Ph with pH (see Section 1.7) or with phenol (C_6H_5OH).

2.3   Alkyl substitution

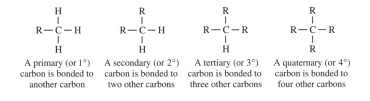

A primary (or 1°) carbon is bonded to another carbon	A secondary (or 2°) carbon is bonded to two other carbons	A tertiary (or 3°) carbon is bonded to three other carbons	A quaternary (or 4°) carbon is bonded to four other carbons

2.4   Naming carbon chains

The IUPAC name of an organic compound is composed of three parts.

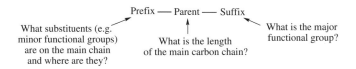

There are four key steps in naming organic compounds.

(1)   Find the longest carbon chain and name this as an alkane. This is the parent name.

No. of carbons	Alkane name	No. of carbons	Alkane name
1	methane	6	hexane
2	ethane	7	heptane
3	propane	8	octane
4	butane	9	nonane
5	pentane	10	decane

(2)   Identify the major functional group. Replace -ane (in the alkane) with a suffix.

> *Functional group priorities*
> carboxylic acid > acid chloride > aldehyde > nitrile >
> ketone > alcohol > amine > halide

major functional group	suffix	major functional group	suffix
alkene	-ene	aldehyde	-anal
alkyne	-yne	ketone	-anone
alcohol	-anol	carboxylic acid	-anoic acid
amine	-ylamine	acid chloride	-anoyl chloride
nitrile	-anenitrile		

(3)   Number the atoms in the main chain. Begin at the end nearer the major functional group and give this the lowest number. For alkanes, begin at the end nearer the first branch point.

(4) Identify the substituents (e.g. minor functional groups) on the main chain and their number. Two substituents on the same carbon are given the same number. The substituent name and position is the prefix. The names of two or more different substituents should be included in alphabetical order in the prefix (e.g. hydroxy before methyl).

minor functional group	prefix	minor functional group	prefix
chloride	chloro-	aldehyde	formyl-
bromide	bromo-	ketone	oxo-
iodide	iodo-	nitro	nitro-
alcohol	hydroxy-	amine	amino-
ether	alkoxy-	nitrile	cyano-

Di- or tri- is used in the prefix or suffix to indicate the presence of two or three of the minor or major functional groups (or substituents), respectively.

Examples

$$\underset{5\quad\ \ 4\quad\ \ 3\quad\ \ 2\quad\ \ 1}{H_3C - CH_2 - \overset{\overset{\displaystyle CH_3}{|}}{CH} - CH_2 - \overset{\overset{\displaystyle O}{\|}}{C} - OH}$$

3-methylpentanoic acid

$$\underset{3\quad\ \ 2\quad\ \ 1}{H_3C - \overset{\overset{\displaystyle Cl}{|}}{CH} - CH_2Cl}$$

1,2-dichloropropane

$$\underset{3\quad\ \ 2\quad\ \ 1}{H_3C - \overset{\overset{\displaystyle OH}{|}}{CH} - CH_2NH_2} \quad \overset{main\ functional}{group}$$

1-amino-2-propanol

$$\underset{5\quad\ \ |4\quad\ \ 3\quad\ \ 2\quad\ \ 1}{H_3C - \overset{\overset{\displaystyle OH}{|}}{\underset{\underset{\displaystyle CH_3}{|}}{C}} - CH_2 - \overset{\overset{\displaystyle O}{\|}}{C} - CH_3} \quad \overset{main\ functional}{group}$$

4-hydroxy-4-methyl-2-pentanone

Special cases

Alkenes and alkynes

The position of the double or triple bond is indicated by the number of the lowest carbon atom in the alkene or alkyne.

$$\underset{5\quad\ \ 4\quad\ \ 3\quad\ \ 2\quad\ \ 1}{H_3C - \overset{\overset{\displaystyle CH_3}{|}}{CH} - CH = CH - CH_3}$$

4-methyl-2-pentene

$$\underset{1\quad\ \ 2\quad\ \ 3\quad\ \ 4}{H_2C = \overset{\overset{\displaystyle CH_3}{|}}{C} - CH = CH_2}$$

2-methyl-1,3-butadiene

$$\overset{major}{\underset{group}{functional}} \quad \underset{1\quad\ \ 2\quad\ \ 3}{H_3C - \overset{\overset{\displaystyle HO}{|}}{\underset{\underset{\displaystyle CH_3}{|}}{C}} - C \equiv CH} \quad \text{2-methyl-3-butyn-2-ol}$$

Aromatics

Monosubstituted derivatives are usually named after benzene (C_6H_6), although some non-systematic or 'trivial' names (in brackets) are still used.

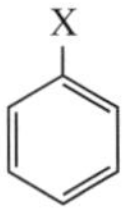

X	Name		X	Name
H	benzene		CH_3	methylbenzene (toluene)
Br	bromobenzene		$CH=CH_2$	ethenylbenzene (styrene)
Cl	chlorobenzene		OH	hydroxybenzene (phenol)
NO_2	nitrobenzene		NH_2	aminobenzene (aniline)
			CN	cyanobenzene (benzonitrile)

The word benzene comes first when functional groups of higher priority (than benzene) are on the ring

X	Name
CHO	benzenecarboxaldehyde (benzaldehyde)
CO_2H	benzenecarboxylic acid (benzoic acid)

Disubstituted derivatives are sometimes named using the prefixes *ortho* (or positions 2- and 6-), *meta* (or positions 3- and 5-) and *para* (or position 4-).

For trisubstituted derivatives, the lowest possible numbers are used and the prefixes are arranged alphabetically.

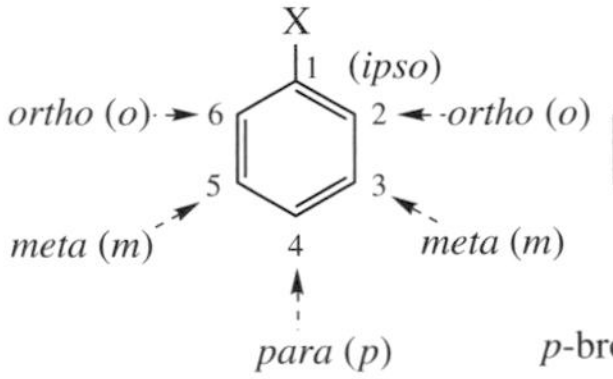

ortho (o) ➤ 6
1 (*ipso*)
2 ◄ *ortho (o)*
5
3
meta (m)
4
meta (m)
para (p)

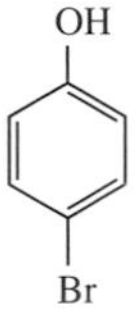

p-bromophenol

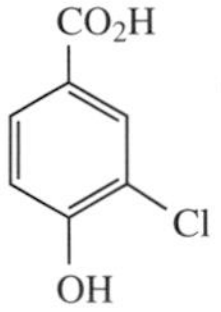

3-chloro-
4-hydroxy-
benzoic acid

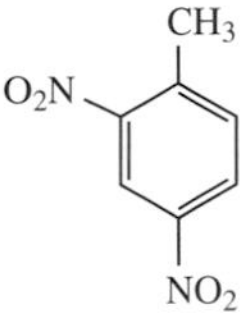

2,4-dinitrotoluene

Esters

These are named in two parts. The first part represents the R^1 group attached to oxygen. The second represents the R^2CO_2 portion which is named as an alkanoate (i.e. the suffix is -anoate; an exception is -oate in benzoate). A space separates the two parts of the name.

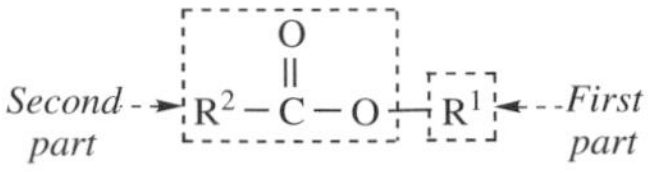

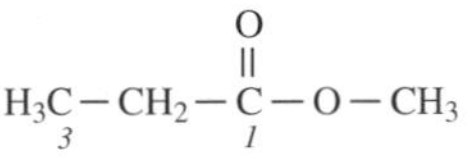

Second part → R^2-C-O R^1 ←-- *First part*

$$H_3C-CH_2-\underset{3}{C}-O-\underset{1}{CH_3}$$

methyl propanoate

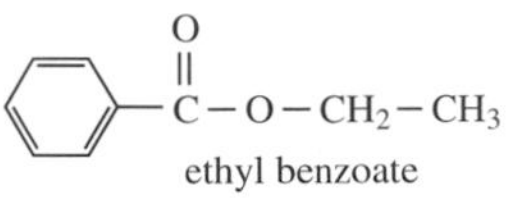

ethyl benzoate

Amides

The R^1 group is the prefix, and $N-$ is written before this to show that the group is on nitrogen.

$$suffix = \atop \text{-anamide} \longrightarrow R^2 - \underset{\overset{\|}{O}}{C} - NH \overset{\cdot\cdot}{\underset{\cdot\cdot}{R^1}} \longleftarrow \text{-}prefix$$

$$H_3\underset{3}{C} - CH_2 - \underset{\overset{\|}{O}}{\underset{1}{C}} - \underset{H}{N} - CH_3$$

N-methylpropanamide

2.5   Drawing organic structures

• In *Kekulé* structures, every carbon atom and every C–H bond are shown.
• In *condensed* structures, the C–H bonds, and often the C–C bonds, are omitted.
• In *skeletal* structures, the carbon and hydrogen atoms are not shown, and the bonds to hydrogen are usually also not shown (although the hydrogen atoms within the functional groups, e.g. alcohols, amines, aldehydes and carboxylic acids, are shown). All other atoms are written. These structures are the most useful (and recommended for use by the reader) because they are uncluttered and quickly drawn, whilst showing all of the important parts of the molecule. Skeletal structures are usually drawn to indicate the approximate shape of the molecule, which is determined by the hybridisation of the atoms (see Section 1.5).

$$H - \overset{\overset{H}{|}}{\underset{\underset{H}{|}}{C}} - \overset{\overset{H}{|}}{\underset{\underset{H}{|}}{C}} - \overset{\overset{O}{\|}}{C} - O - \overset{\overset{H}{|}}{\underset{\underset{H}{|}}{C}} - H \quad \equiv \quad CH_3CH_2CO_2CH_3 \quad \equiv$$

Kekulé   condensed   skeletal

$$H - \overset{\overset{H}{|}}{\underset{\underset{H}{|}}{C}} - \overset{H}{C} = \overset{H}{C} - \overset{\overset{H}{|}}{\underset{\underset{H}{|}}{C}} - OH \quad \equiv \quad CH_3CH=CHCH_2OH \quad \equiv \quad \text{/} \diagdown \text{/} \diagdown OH$$

Kekulé   condensed   skeletal

Benzene can be written with a circle within the ring to show the delocalisation of electrons (see Section 7.1). However, this does not show the six π-electrons, which makes drawing reaction mechanisms impossible. A single resonance form, showing the three C=C bonds, is therefore most often used.

Problems

(1) Draw structures for each of the following compounds.
   (a) 1-Bromo-4-chloro-2-nitrobenzene
   (b) Methyl-3-bromobutanoate
   (c) N-Methylphenylethanamide
   (d) 2-(3-Oxobutyl)cyclohexanone
   (e) 4-Hexen-2-one
   (f) 2-Buten-1-ol
   (g) 6-Chloro-2,3-dimethyl-2-hexene
   (h) 1,2,3-Trimethoxypropane
   (i) 2,3-Dihydroxybutanedioic acid (tartaric acid)
   (j) 5-Methyl-4-hexenal

(2) Name the following compounds.

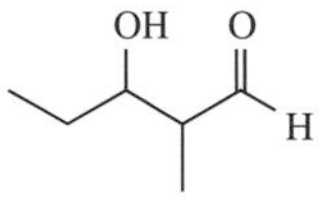

(a) 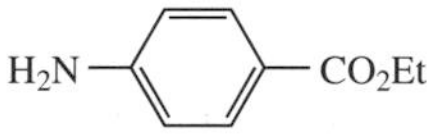   (b) 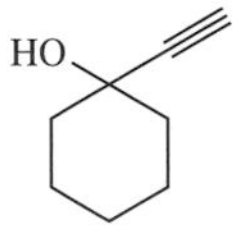   (c)

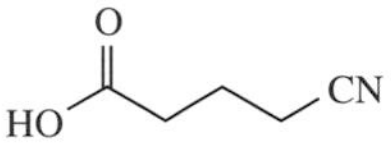

(d) 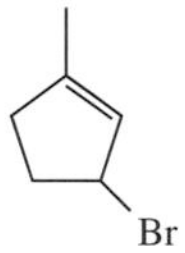   (e) 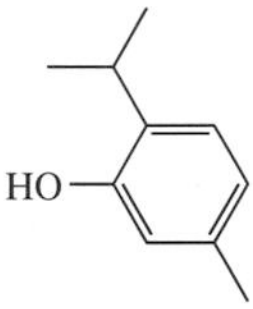   (f)

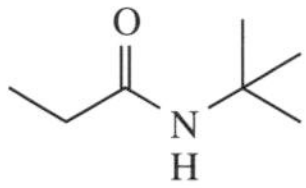

(g) 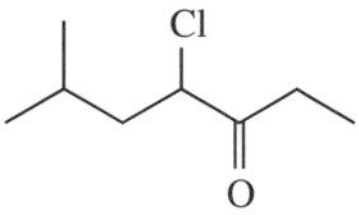   (h)

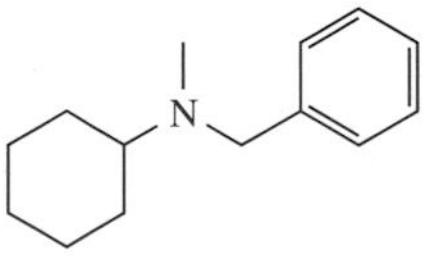

(i) 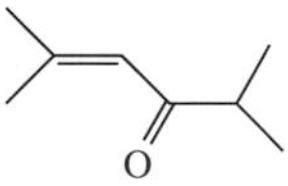   (j)

3. STEREOCHEMISTRY

Key point. The spatial arrangement of atoms determines the *stereochemistry*, or shape, of organic molecules. When different shapes of the same molecule are interconvertible on rotating a bond, they are known as *conformational isomers*. In contrast, *configurational isomers* cannot be interconverted without breaking a bond, and examples include alkenes and *optical isomers*, which rotate plane-polarised light.

3.1 Isomerism

Isomers are compounds that have the same numbers and the same kinds of atoms, but they differ in the way the atoms are arranged.

- *Constitutional isomers* are compounds which have the same molecular formula but have the atoms joined together in a different way. They have different physical and chemical properties.

Examples

CH_3 $\|$ $H_3C - C - CH_3$ $\|$ H 2-methylpropane (branched chain)	and $\quad H_3C - CH_2 - CH_2 - CH_3$ butane (straight chain)	Both C_4H_{10} but different carbon skeleton
OH $\|$ $H_3C - C - CH_3$ $\|$ H 2-propanol (secondary alcohol)	and $\quad H_3C - CH_2 - CH_2 - OH$ 1-propanol (primary alcohol)	Both C_3H_8O but different position of the functional group
OH $\|$ $H_3C - C - CH_3$ $\|$ CH_3 2-methyl-2-propanol (tertiary alcohol)	and $H_3C - CH_2 - O - CH_2 - CH_3$ diethyl ether (ether)	Both $C_4H_{10}O$ but different functional groups

- *Conformational isomers* are different shapes of the same molecule resulting from rotation around a single C–C bond. They are not different compounds (i.e. they have the same physical and chemical properties) and are readily interconvertible (see Section 3.2).

- *Configurational isomers* have the same molecular formula, and although the atoms are joined together in the same way, they are arranged differently in space (with respect to each other). They are not readily interconvertible (see Section 3.3).

3.2 Conformational isomers

The different arrangements of atoms caused by rotation about a single bond are called conformations. A *conformer* (or conformational isomer) is a compound with a particular conformation. Conformational isomers can be represented by *Sawhorse representations* or *Newman projections*.

3.2.1 Conformations of ethane (CH_3CH_3)

Rotation around the C–C bond produces two distinctive conformations.

- *Eclipsed conformation.* C–H bonds on each carbon atom are as close as possible.

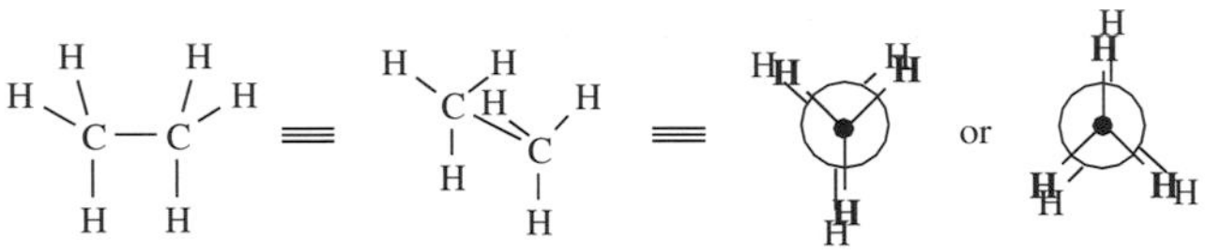

Sawhorse projection
(view the C–C bond from an oblique angle and show all C–H bonds)

Newman projection
(view the C–C bond directly on, show the front carbon atom by a small black circle and the back carbon by a much larger circle. Show all C–H bonds)

- *Staggered conformation.* C–H bonds on each carbon atom are as far apart as possible.

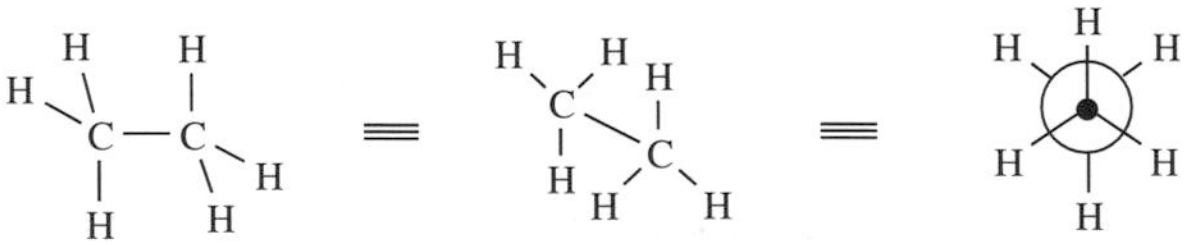

Sawhorse projection   Newman projection

The staggered conformation is more stable as the C–H bonds are further apart. The energy difference between them ($12\,kJ\,mol^{-1}$) is known as the *torsional strain*. This energy difference is relatively small, and hence there is free rotation about the C–C bond at room temperature.

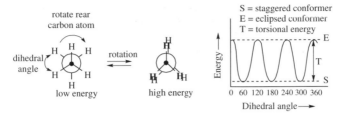

The angle between the C–H bonds on the front and back carbon is known as the *dihedral* (or *torsional*) *angle*.

3.2.2   Conformations of butane ($CH_3CH_2CH_2CH_3$)

In butane, the staggered conformations do not all have the same energy.

- The *anti-* (or *antiperiplanar*) *conformation* is the most stable as the two methyl groups are as far apart as possible (180° separation).
- The *gauche* (or *synclinal*) *conformation* is higher in energy (by $4\,kJ\,mol^{-1}$) as the two methyl groups are near one another (60° separation), resulting in *steric strain*. Steric strain is the repulsive interaction between two groups, which are closer to one another than their atomic radii allow.

anti- conformation
(more stable)

gauche conformation
(less stable)

The most stable conformation for a straight-chain alkane is a zigzag shape, as the alkyl groups are as far apart as possible. (This is why we draw alkyl chains with a zigzag structure.)

zigzag shape
of butane

It should be noted that when the methyl groups in butane are eclipsed and are in the same plane (0° separation), this is known as the *synperiplanar* conformation.

3.2.3   Conformations of cycloalkanes

The shape of cycloalkanes is determined by torsional strain, steric strain and angle strain.

- *Angle strain.* For an sp^3 hybridised carbon atom, the ideal bond angle is 109.5° (tetrahedral). Angle strain is the extra energy that a compound has because of non-ideal bond angles (i.e. angles above or below 109.5°).

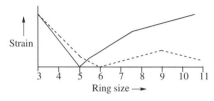

——— = **Angle strain**

A 3-ring has the highest and a planar 5-ring has the lowest angle strain. After 5, the angle strain increases as the ring gets larger

- - - - = **Total strain**
(angle + steric + torsional strain)

A 3-ring has the highest total strain which reaches a minimum for a 6-ring. The strain increases from 6 to 9 and then decreases

Cyclopentane (5-membered) and cyclohexane (6-membered) rings are therefore the most stable and, consequently, the most easily formed.

Cycloalkanes can adopt different conformations (or shapes): cyclopropane is flat, cyclobutane can form a butterfly shape, while cyclopentane can form an open-envelope shape.

cyclopropane
(internal C–C
angle = 60°)

cyclobutane
(butterfly)

cyclopentane
(open-envelope)

Cyclopropane has to be planar and therefore has very strained bond angles of 60° and a great deal of torsional energy. However, cyclobutane and cyclopentane can adopt non-planar (puckered) shapes which decrease the torsional strain by staggering the C–H bonds. However, this is at the expense of angle strain, and the butterfly and open-envelope shapes represent the best compromise between the two opposing effects.

3.2.4 Cyclohexane

Cyclohexane adopts the chair or boat conformations, which are both free of angle strain. However, the boat conformation is less stable because of steric strain between the C-1 and C-4 (or flagstaff) hydrogens. The two chair forms can interconvert via the boat form in a process known as *ring-flipping*.

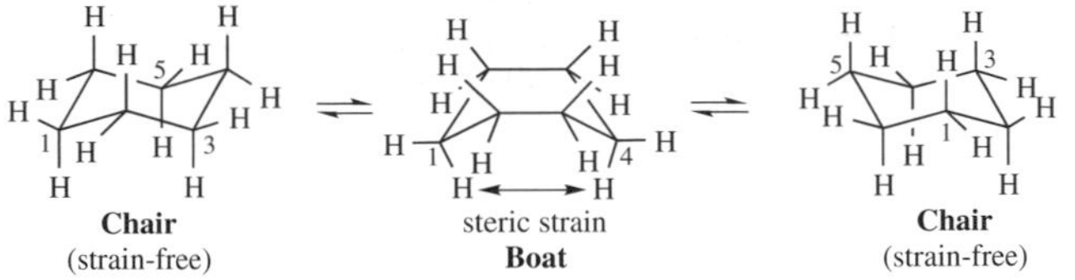

| Chair | Boat | Chair |
| (strain-free) | steric strain | (strain-free) |

Newman projections

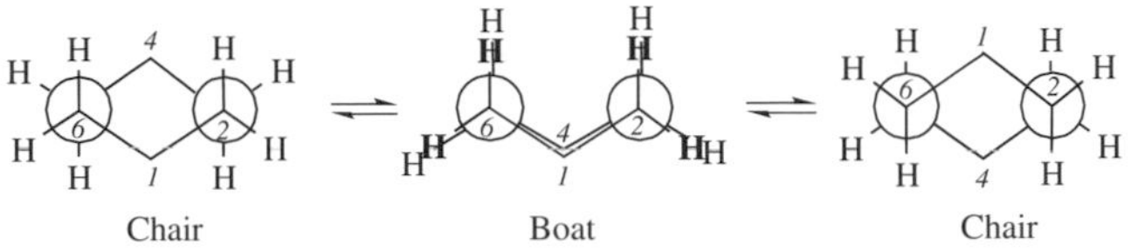

| Chair | Boat | Chair |

The chair conformation has six *axial* and six *equatorial* hydrogens. On ring-flipping, the axial hydrogens become equatorial and the equatorial hydrogens become axial.

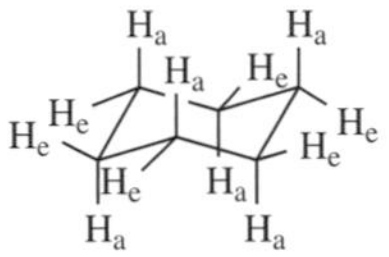

H_a = axial hydrogen (point up and down)

H_e = equatorial hydrogen (point sideways)

If a substituent (X) is present, then this prefers to sit in an *equatorial position*. The equatorial conformer is lower in energy because *steric strain* (or *1,3-diaxial interactions*) raises the energy of the axial conformer. As the size of the X group increases, so does the proportion of the equatorial conformer at equilibrium.

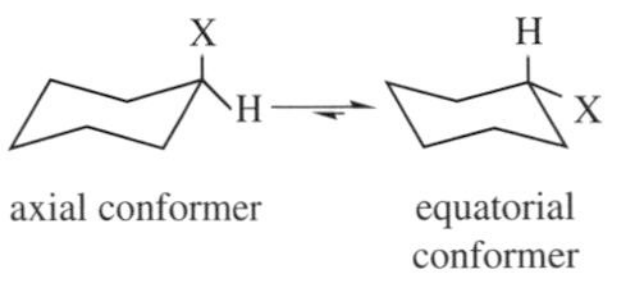

X	% eq. : % ax.	
Me	95	5
^{t}Bu	>99	<1

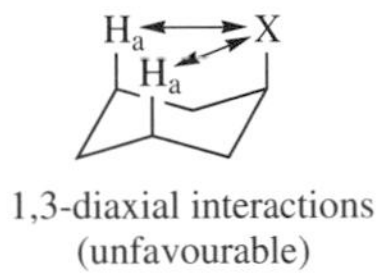

axial conformer   equatorial conformer

1,3-diaxial interactions (unfavourable)

For disubstituted cyclohexanes, both groups should sit in an equatorial position. When this is not possible, then the largest group (e.g. a ^{t}Bu group rather than a Me group) will sit in the equatorial position.

diaxial conformer   diequatorial conformer

The bulkier ^{t}Bu group prefers the equatorial position

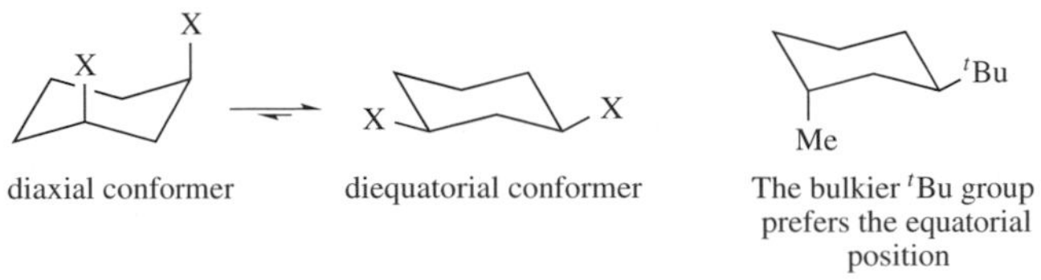

When the two substituents on the ring are *both* pointing up (or *both* pointing down), these cyclic compounds are designated *cis*-stereoisomers. When one substituent is pointing up and the other down, these cyclic compounds are designated *trans*-stereoisomers (see Section 3.3).

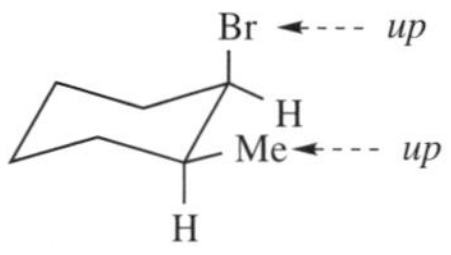

cis-1-bromo-2-methylcyclohexane 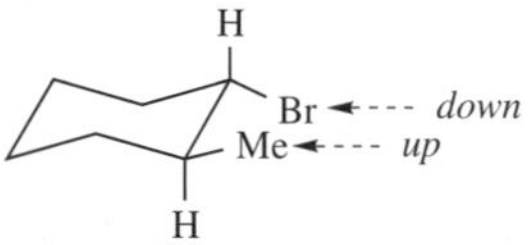   trans-1-bromo-2-methylcyclohexane

3.3 Configurational isomers

The spatial arrangement of atoms or groups in molecules is known as *configuration*. Compounds with the same molecular formula and the same types of bond, which cannot be interconverted without breaking a bond, have different configurations (and are configurational isomers).

3.3.1 Alkenes

Alkenes with two different substituents (A, B, D or E) at each end of the double bond can exist as two configurational isomers because there is no rotation around the C=C bond.

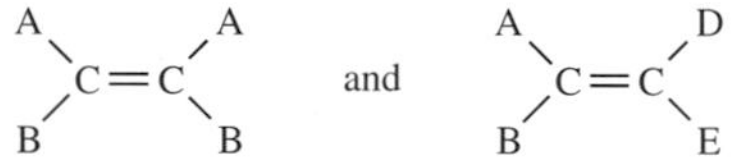

can have configurational isomers

3.3.1.1 *Cis*- and *trans*-isomerism

For disubstituted alkenes (with two substituents on the double bond), alkenes can be named using the *cis–trans* nomenclature:

- *cis*-isomers have the substituents on the same side
- *trans*-isomers have the substituents on the opposite side.

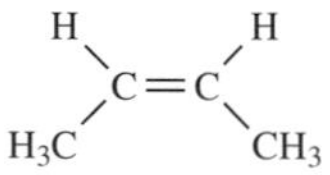

cis-2-butene 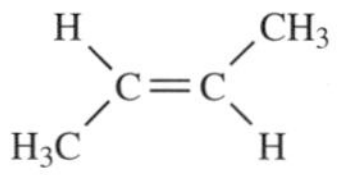   trans-2-butene 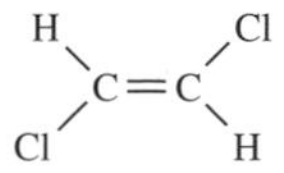   trans-1,2-dichloroethene

Cis- and *trans*-isomers are also called geometric isomers because they have different shapes.

There are no rigid rules for deciding whether a variety of substituted double bonds are *cis*- or *trans*-, and hence the more systematic *E/Z* nomenclature is generally preferred (see Section 3.3.1.2).

3.3.1.2   *E* and *Z* nomenclature

For di-, tri- and tetra-substituted alkenes (with two, three and four substituents, respectively, on the double bond), alkenes can be named using the E,Z nomenclature. The groups on the double bond are assigned priorities based on a series of sequence rules.

- *E*-alkenes have the groups of highest priority on the *opposite* sides
- *Z*-alkenes have the groups of highest priority on the *same* sides.

Sequence rules

(1) Rank the atoms directly attached to the double bond carbons in the order of decreasing atomic number. The highest atomic number is ranked first.

$$Atom = Br > Cl > O > N > C > H$$
$$Atomic\ number = 35 > 17 > 8 > 7 > 6 > 1$$

(2) If the atoms directly linked to the double bond are the same, then the second, third, fourth, etc. atoms (away from the double bond) are ranked until a difference is found.

$$-\xi-CH_2-CH_3 \quad > \quad -\xi-CH_3 \qquad\qquad -\xi-O-CH_3 \quad > \quad -\xi-OH$$

   ethyl   methyl   methoxy   hydroxy

(3) Multiple (double or triple) bonds are assumed to have the same number of single-bonded atoms.

$$\text{aldehyde}$$

$$\begin{array}{c} H \\ \diagdown \\ C=O \end{array} \equiv \begin{array}{c} H \quad O \quad C \\ \diagdown | \quad | \\ C-O \end{array} \qquad\qquad \text{nitrile}$$

$$-\xi-C\equiv N \quad \equiv \quad -\xi-\underset{N}{\overset{N}{\underset{|}{\overset{|}{C}}}}-\underset{C}{\overset{C}{\underset{|}{\overset{|}{N}}}}$$

Assume carbon is bonded to two oxygens   Assume carbon is bonded to three nitrogens
Assume oxygen is bonded to two carbons   Assume nitrogen is bonded to three carbons

Examples (numbers on each carbon are given in italic to indicate priorities)

$$\begin{array}{cc} \overset{1}{Cl} & \overset{1}{CH_2CH_3} \\ \diagdown\quad\diagup \\ C=C \\ \diagup\quad\diagdown \\ \overset{2}{H_3C} & \overset{2}{CH_3} \end{array}$$

(*Z*)-2-chloro-3-methyl-2-pentene

$$\begin{array}{cc} N & \\ \diagdown\diagdown & \\ \overset{2}{C} & \overset{1}{CH_2OH} \\ \diagdown\quad\diagup \\ C=C \\ \diagup\quad\diagdown \\ \overset{1}{Br} & \overset{2}{CH_2CH_3} \end{array}$$

(*E*)-2-bromo-3-hydroxymethyl-2-pentenenitrile

$$\begin{array}{cc} O & \\ \| & \\ C & \overset{2}{CH_3} \\ H_3C \diagup \overset{1}{} \diagdown\quad\diagup \\ C=C \\ \diagup\quad\diagdown \\ \overset{2}{H_3C} & \overset{1}{C_6H_5} \end{array}$$

(*E*)-3-methyl-4-phenyl-3-penten-2-one

$$\begin{array}{cc} \overset{1}{HO_2C} & \overset{1}{CHO} \\ \diagdown\quad\diagup \\ C=C \\ \diagup\quad\diagdown \\ \overset{2}{Ph} & \overset{2}{CH_2OH} \end{array}$$

(*Z*)-3-hydroxymethyl-4-oxo-2-phenylbut-2-enoic acid

3.3.2   Optical isomers

- *Optical isomers* are configurational isomers with the same chemical and physical properties, which are able to rotate plane-polarised light clockwise or anticlockwise.
- Asymmetrical molecules, which are non-identical with their mirror images, are known as *chiral* molecules, and these can rotate plane-polarised light. If a molecule has a plane of symmetry, it cannot be chiral, and it is known as an *achiral* molecule.
- Molecules with a (asymmetric) tetrahedral carbon atom bearing four different groups are chiral. The 3-dimensional structure of the molecule, showing the position of the groups attached to the tetrahedral carbon atom, can be represented using solid/dashed wedges in *flying wedge formulae.*

$$R^1$$
$$|$$
$$C\,{,,,,}\,R^3$$
$$R^2 \diagup \quad R^4$$

— solid wedge, the bond points toward the reader

⋯⋯⋯ dashed wedge, the bond points away from the reader

— single line, shows the (two) bonds in the plane of the paper.

3.3.2.1   Enantiomers

The tetrahedral asymmetric carbon atom is known as a *chiral* or *stereogenic centre*. These molecules, which are non-identical (not superimposable) with their mirror images, are called *enantiomers*.

mirror
plane

$$R^1 \qquad\qquad R^1$$
$$| \qquad\qquad\qquad |$$
$$C\,{,,,,}\,R^3 \qquad R^3\,{,,,,}\,C$$
$$R^2 \diagup \quad R^4 \qquad R^4 \diagup \quad R^2$$

When $R^1 \neq R^2 \neq R^3 \neq R^4$, these two mirror images are not superimposable – they are **enantiomers**

Enantiomers rotate plane-polarised light in the opposite directions. The (+)-enantiomer (or dextrorotatory enantiomer) rotates the light to the right, while the (−)-enantiomer (or laevorotatory enantiomer) rotates the light to the left. The amount of rotation is called the *specific rotation*, $[\alpha]_D$, and this is measured using a polarimeter. (By convention, the units of $[\alpha]_D$ are often not quoted.)

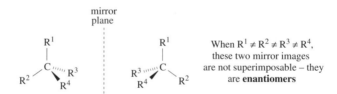

$$[\alpha]_D^T \;=\; \frac{\text{observed rotation (degrees)} \times 100}{\text{path length (dm)} \times \text{concentration (g per 100 cm}^3)} \;=\; \frac{\alpha \times 100}{l \times c}$$

$(10^{-1}\ \text{degree cm}^2\ \text{g}^{-1})$

(where D = sodium D line, i.e. light of $\lambda = 589$ nm; T = temperature in °C)

A 1:1 mixture of two enantiomers is known as a *racemate* or *racemic mixture*, and this does not rotate plane-polarised light. The separation of a racemic mixture into its two enantiomers is called *resolution*.

Example: 2-hydroxypropanoic acid (lactic acid)

mirror
plane

H   H
| |
C''''''CH₃   H₃C''''''C
HO₂C   OH   HO   CO₂H

(+)-enantiomer   (−)-enantiomer
$[\alpha]_D^{25} = +3.82$ $[\alpha]_D^{25} = -3.82$

The optical purity of a compound is a measure of the enantiomeric purity, and this is described as the *enantiomeric excess* (*ee*).

$$ee = \text{per cent of one enantiomer} - \text{per cent of the other enantiomer} = \frac{\text{observed } [\alpha]_D}{[\alpha]_D \text{ of pure enantiomer}} \times 100$$

For example, a 50% *ee* corresponds to a mixture of 75% of one enantiomer and 25% of the other enantiomer.

3.3.2.2   The Cahn–Ingold–Prelog (*R* and *S*) nomenclature

The 3-dimensional arrangement of atoms attached to a stereogenic (or asymmetric) carbon centre is known as the *configuration*. The configuration can be assigned *R* or *S* using the following *Cahn–Ingold–Prelog sequence rules*. This uses the same rules as for *E* and *Z* nomenclature (see Section 3.3.1.2)

(1) Rank the atoms directly attached to the stereogenic carbon atom in the order of decreasing atomic number. The highest atomic number is ranked first.

(2) If the atoms directly linked to the stereogenic centre are the same, then the second, third, fourth, etc. atoms (away from the stereogenic centre) are ranked until a difference is found.

(3) Multiple (double or triple) bonds are assumed to have the same number of single-bonded atoms.

(4) The molecule is then orientated so that the group of the lowest (fourth) priority is drawn pointing away from the reader. (Be careful not to swap the relative positions of the groups around.) The lowest priority group can now be ignored. A curved arrow is drawn from the highest priority group to the second and then to the third highest priority group. (If you have to draw a certain stereoisomer, always start with the lowest priority group pointing away from you.)

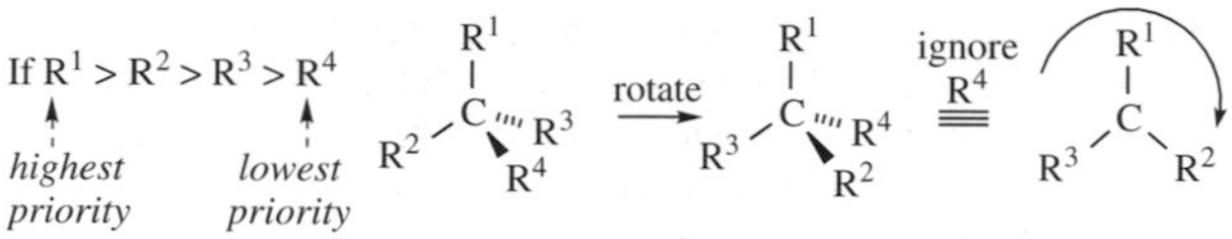

If $R^1 > R^2 > R^3 > R^4$

highest priority (↑ at R^1)   lowest priority (↑ at R^4)

(5) If the arrow is clockwise, then the stereogenic centre has the *R* configuration. If the arrow is anticlockwise, then the stereogenic centre has the *S* configuration. (The assignment of *R* may be remembered by analogy to a car steering wheel making a *right* turn.)

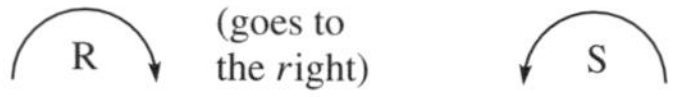

R (goes to the *r*ight)   S

Examples (numbers on each carbon are given in italic to indicate priorities)

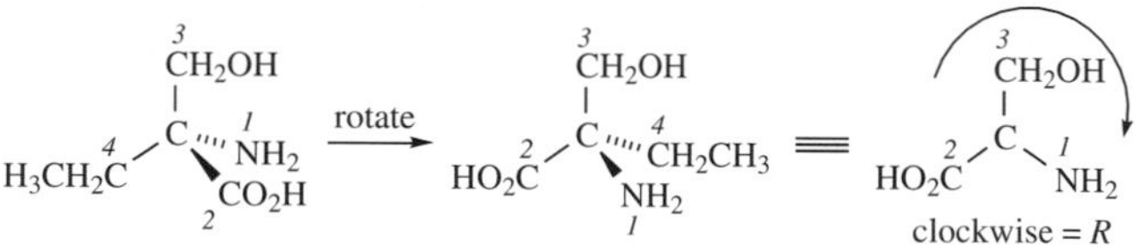

(*R*)-2-amino-2-hydroxymethyl-butanoic acid

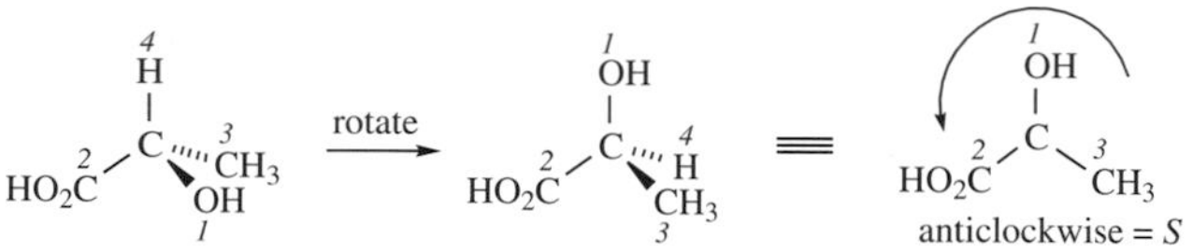

clockwise = *R*

(*S*)-2-hydroxypropanoic acid or (*S*)-lactic acid

anticlockwise = *S*

It just so happens that (*S*)-lactic acid is the same as (+)-lactic acid (see Section 3.3.2.1). However, the + or − sign of the optical rotation is not related to the *R*,*S* nomenclature (i.e. a (+)-enantiomer could be either an *R*-enantiomer or an *S*-enantiomer).

3.3.2.3   The D and L nomenclature

This old-fashioned nomenclature uses glyceraldehyde (or 2,3-dihydroxypropanal) as the standard. The (+)-enantiomer is given the label D (from *d*extrorotatory), and the (−)-enantiomer is given the label L (from *l*aevorotatory).

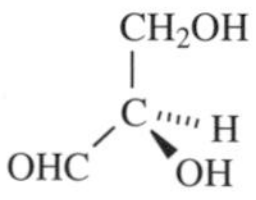

D-(+)-glyceraldehyde 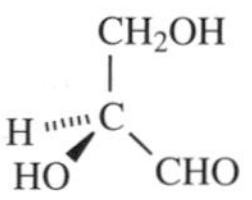   L-(−)-glyceraldehyde

Any enantiomerically pure compound that is prepared from, for example, D-glyceraldehyde is given the label D. Any enantiomerically pure compound that can be converted to, for example, D-glyceraldehyde is given the label D.

D-Glyceraldehyde is the same as (R)-glyceraldehyde. However, the D or L label is not related to the R,S nomenclature (i.e. a D-enantiomer could be either an R-enantiomer or an S-enantiomer). This nomenclature is still used to assign natural products including sugars and amino acids (see Sections 11.1 and 11.3).

3.3.2.4 Diastereoisomers (or diastereomers)

For compounds with two stereogenic centres, four stereoisomers are possible, as there are four possible combinations of R and S.

- *Diastereoisomers* (or diastereomers) are stereoisomers that are not mirror images of each other. This means that diastereoisomers must have a different (R or S) configuration at *one* of the two chiral centres.
- *Enantiomers* must have a different (R or S) configuration at *both* chiral centres.

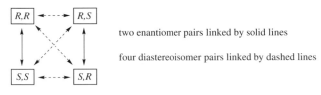

two enantiomer pairs linked by solid lines

four diastereoisomer pairs linked by dashed lines

Example (the numbers represent the numbering of the carbon backbone)

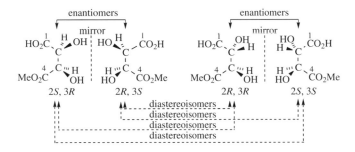

For a compound containing n chiral centres, the total number of stereoisomers will be 2^n and the number of pairs of enantiomers will be 2^{n-1}.

3.3.2.5 Diastereoisomers versus enantiomers

- Enantiomers have identical chemical and physical properties, except for their biological activity (i.e. they interact differently with other chiral molecules) and their effects on plane-polarised light.

- Diastereoisomers can have different physical (e.g. melting point, polarity) and chemical properties.
- Enantiomers are always chiral.
- Diastereoisomers can be chiral or achiral. If a diastereomer contains a plane of symmetry, it will be achiral. These compounds, which contain stereogenic centres but are achiral, are called *meso compounds*. In effect, the plane of symmetry divides the molecule into halves, which contribute equally but opposite to the rotation of plane-polarised light (i.e. they cancel each other out).

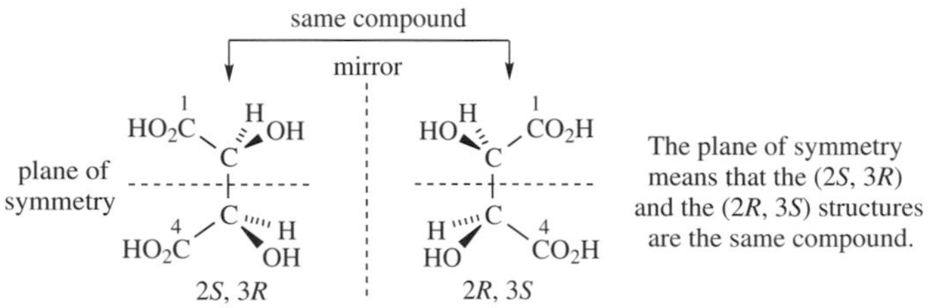

The plane of symmetry means that the (2S, 3R) and the (2R, 3S) structures are the same compound.

meso-tartaric acid

3.3.2.6   Fischer projections

Molecules with asymmetric carbons can be represented by Fischer projections, in which a tetrahedral carbon atom is represented by two crossed lines. The vertical lines represent the bonds pointing into the page (away from the reader), and the horizontal lines represent the bonds pointing out of the page (towards the reader).

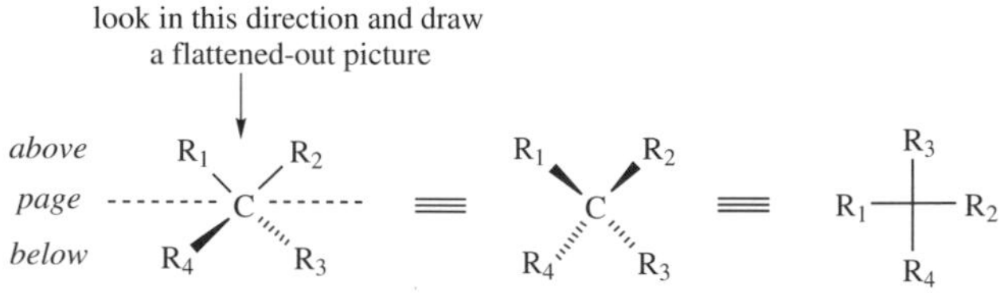

The compound is usually drawn so that the main carbon skeleton is vertical and the highest priority functional group is at the top of the vertical line.

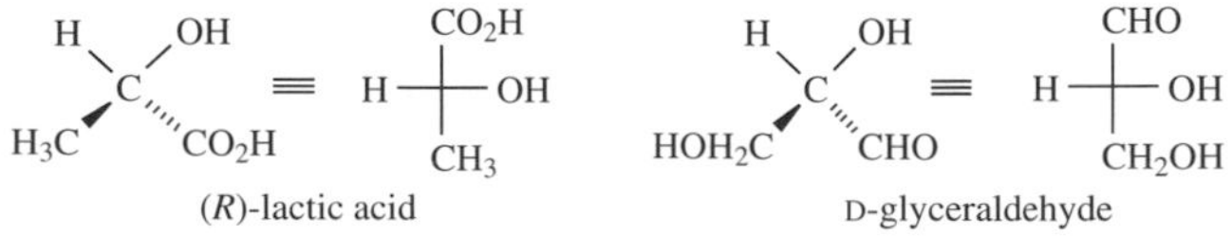

This old-fashioned nomenclature is generally only used to show amino acids and sugars, which can contain several asymmetric centres (see Sections 11.1 and 11.3).

Example: D-*erythrose or (2R,3R)-trihydroxybutanal*

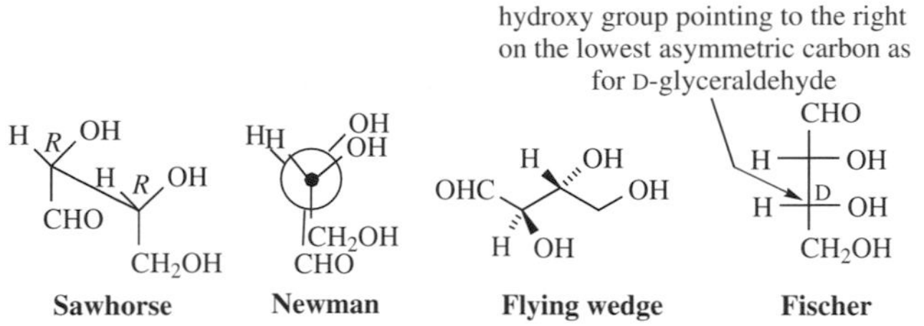

hydroxy group pointing to the right
on the lowest asymmetric carbon as
for D-glyceraldehyde

| Sawhorse | Newman | Flying wedge | Fischer |

The *R,S* nomenclature can be assigned to Fischer projections by drawing the substituent of lowest priority in a vertical position (i.e. at the top). Although Fischer projections can be rotated in the plane of the paper by 180°, changing the position of substituents by 90° requires 'double exchanges'.

Examples (numbers on each carbon are given in italic to indicate priorities)

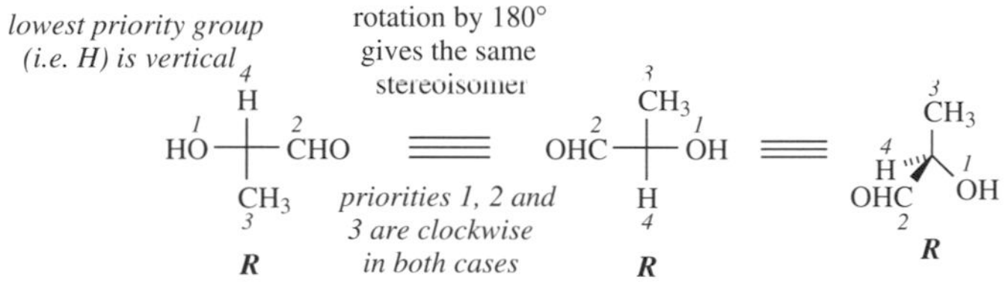

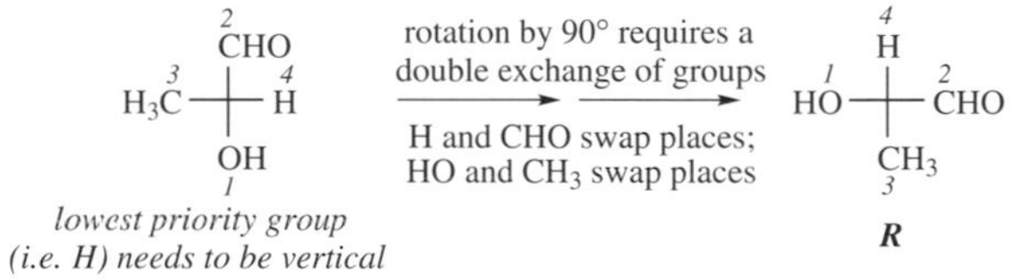

Problems

(1) Consider the stereochemical priorities of the following groups using the Cahn–Ingold–Prelog system.

$-CONH_2$, $-CH_3$, $-CO_2H$, $-CH_2Br$, $-I$, $-CCl_3$, $-OCH_3$

(a) Which group has the highest and which has the lowest priority?
(b) Which group ranks between $-CCl_3$ and $-CONH_2$?
(c) Assign prefixes (Z) and (E) as appropriate to compounds **A**, **B** and **C**.

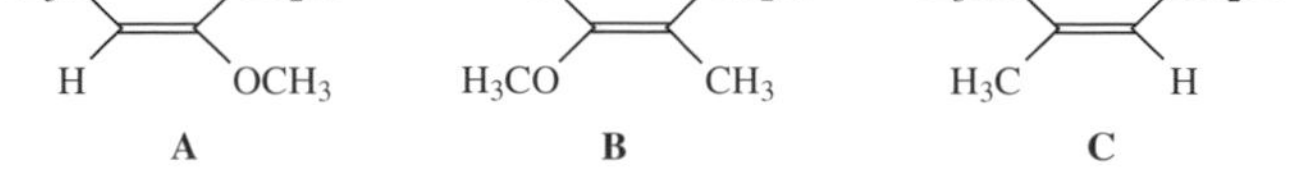

| A | B | C |

(2) Assign the configuration (R or S) to the compounds (**D–F**) shown below.

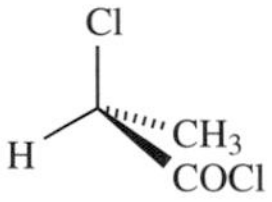

D 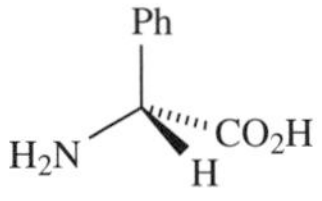   **E** 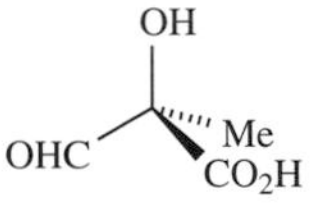   **F**

(3) The Fischer projection of *meso*-2,3-dihydroxybutane (**G**) is shown below.
   (a) Why is this a *meso* compound?
   (b) Draw Newman and Sawhorse projections of the antiperiplanar conformation of **G**.

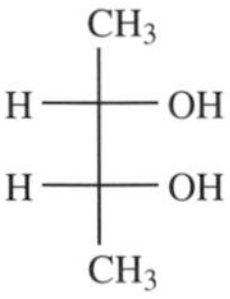

G

   (c) Draw a Fischer projection of a diastereoisomer of **G**.

(4) Draw the preferred chair conformation of ($1R,2S,5R$)-(−)-menthol (**H**).

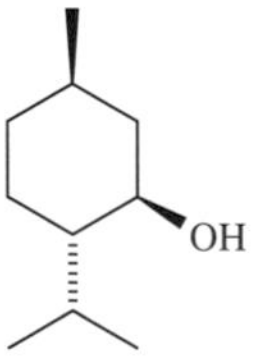

H

(5) What are the stereochemical relations (identical, enantiomers, diastereoisomers) of the following four molecules (**I–L**)? Assign absolute configurations at each stereogenic centre.

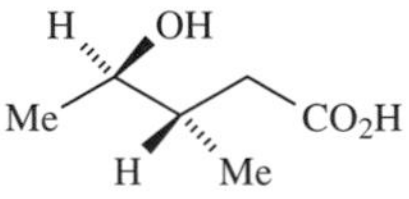

I 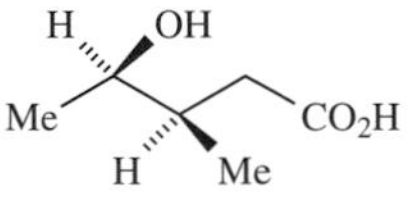   **J**

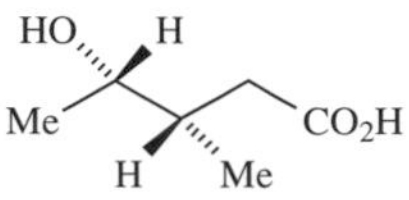

K 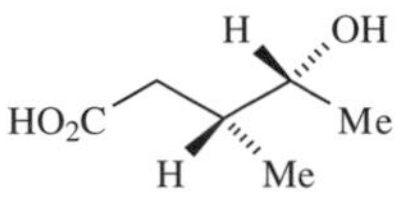   **L**

(6) What are the stereochemical relations (identical, enantiomers, diastereoisomers) of the following four molecules (**M–P**)? Assign absolute configurations at each stereogenic centre.

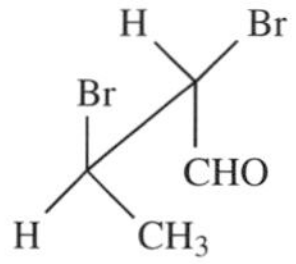

M

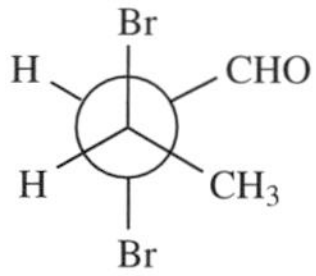

N

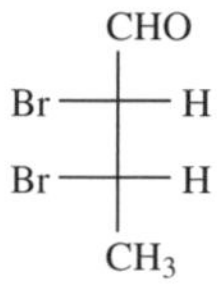

O

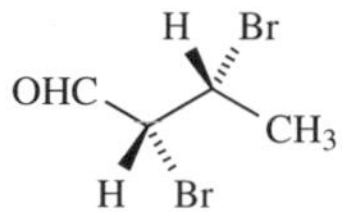

P

4. REACTIVITY AND MECHANISM

Key point. Organic reactions can take place by *radical* or, more commonly, by *ionic* mechanisms. The particular pathway of a reaction is influenced by the stability of the intermediate radicals or ions, which can be determined from an understanding of *electronic* and *steric* effects. For ionic reactions, *nucleophiles* (electron-rich molecules) form bonds to *electrophiles* (electron-deficient molecules), and this can be represented using curly arrows. The energy changes that occur during a reaction can be described by the *equilibria* (i.e. how much of the reaction occurs) and also by the *rate* (i.e. how fast the reaction occurs). The position of the equilibrium is determined by the size of the *Gibbs free energy change*, while the rate of a reaction is determined by the *activation energy*.

4.1 Reactive intermediates: ions versus radicals

There are two ways of breaking a covalent bond. The unsymmetrical cleavage is called *heterolytic cleavage* (or heterolysis), and this leads to the formation of ions (cations and anions). The symmetrical cleavage is called *homolytic cleavage* (or homolysis), and this leads to the formation of radicals.

Curly arrows can be used to represent bond cleavage. A double-headed arrow represents the movement of two electrons (and is used in ionic mechanisms). A single-headed arrow (or fishhook) is used to represent the movement of a single electron (and is used in radical mechanisms). Curly arrows therefore always depict the movement of electrons.

Heterolysis (use one double-headed arrow)

$$A\!-\!B \longrightarrow A^{\oplus} + B^{\ominus}$$

cation   anion

Ions contain an even number of electrons

Both electrons in the two-electron bond move to only one atom

Homolysis (use two single-headed arrows)

$$A\!-\!B \longrightarrow A^{\bullet} + B^{\bullet}$$

radical   radical

The two-electron bond is split evenly and one electron moves to each of the atoms

Radicals contain an odd number of electrons, and a dot is used to represent the unpaired electron.

Processes that involve unsymmetrical bond cleavage or bond formation are known as *ionic* (or *polar*) *reactions*. Processes that involve symmetrical bond cleavage or bond formation are known as *radical reactions*.

Heterolytic bond formation (the reverse of heterolysis)

A two-electron arrow goes from the anion/lone pair to the cation. The arrow points from the electron-rich centre to the electron-deficient centre

Homolytic bond formation (the reverse of homolysis)

The one-electron arrows point towards one another to make a new two-electron bond. The arrow points away from the dot and finishes midway between A and B

Formally, the two-electron arrow should point to where the new bond will be formed if the electrons are being used to form the bond. If the electrons end up as a lone pair, then they point to the atom. (Many textbooks show arrows drawn directly onto the atom in both cases.)

When more than one two-electron arrow is used in a reaction scheme, the arrows *must* always point in the same direction.

For any reaction, the overall charge of the starting materials should be the *same* as that of the product(s).

4.2   Nucleophiles and electrophiles

- *Nucleophiles* are electron-rich species that can form a covalent bond by *donating* two electrons to an electron-deficient site. Nucleophiles are negatively charged (anions) or neutral molecules that contain a lone pair of electrons.
- *Electrophiles* are electron-deficient species that can form a covalent bond by *accepting* two electrons from an electron-rich site. Electrophiles are often positively charged (cations), although they can also be neutral.

The nucleophilic or electrophilic sites within a neutral organic molecule can be determined by: (i) the presence of lone pairs of electrons;

(ii) the type of bonding (sp, sp^2 or sp^3); and/or (iii) the polarity of the bonds.

(1)   An atom (such as nitrogen, oxygen or sulfur) bearing an electron pair will be a nucleophilic site.
(2)   Double or triple carbon–carbon bonds in alkenes, alkynes and aromatics are of high electron density and hence are nucleophilic sites. (Single C–C bonds in alkanes are not nucleophilic.)
(3)   In a polar bond (see Section 1.6.1), the electrons are held closer to the more electronegative atom. The electronegative atom will be a nucleophilic site, and the less electronegative atom will be an electrophilic site.

Single bonds

Y is *more* electronegative than carbon	Y is *less* electronegative than carbon
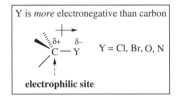	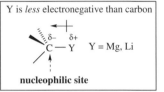

Double and triple bonds

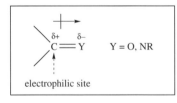

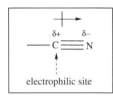

4.2.1   Relative strength

4.2.1.1   Nucleophiles

A nucleophile with a negative charge is always a more powerful nucleophile than its conjugate acid (which is neutral).

$$\text{Nucleophilic strength} \qquad HO^{\ominus} \; > \; H_2\ddot{O}$$

The relative nucleophilic strength (or nucleophilicity) of an anion, or a nucleophilic site within a neutral molecule, depends on the availability of the two electrons. The more electronegative the atom, the less nucleophilic the atom will be, because the electrons are held tighter to the nucleus.

The nucleophilic strength of anions, within the same row of the periodic table, follows the same order as basicity: the more electronegative the atom bearing the negative charge, the weaker the nucleophile and the weaker the base.

Nucleophilic strength $\qquad$ $R_3C^{\ominus} > R_2N^{\ominus} > RO^{\ominus} > F^{\ominus}$
(or basicity)

Electronegativity $\qquad$ $C < N < O < F$

Nucleophilic strength $\qquad$ $R-\overset{..}{N}H_2 > R-\overset{..}{O}H > R-\overset{..}{F}$
(or basicity)

Basicity = donation of a pair of electrons to H (or H$^+$)	**Nucleophilicity** = donation of a pair of electrons to an atom other than H

The pK_a-values can therefore be used to estimate the nucleophilicity of atoms within the same row. It is not exact because the nucleophilicity is strongly affected by steric factors (see Section 4.4), while, the basicity is not.

The nucleophilic strength of anions and neutral atoms increases on going down a group of the periodic table. The electrons are held less tightly to the nucleus as the atom size increases, and hence they are more available for forming bonds. Larger atoms, with more loosely held electrons (than smaller atoms), are said to have a higher *polarisability*.

Nucleophilic strength $\qquad$ $H_2\overset{..}{S} > H_2\overset{..}{O}$

$I^{\ominus} > Br^{\ominus} > Cl^{\ominus} > F^{\ominus}$

The nucleophilic strength of anions depends on the solvent.

- Anions are generally more nucleophilic in *aprotic solvents* (these contain polar groups but no O–H or N–H bonds), such as dimethyl sulfoxide (Me$_2$SO), than in protic solvents.
- In *protic solvents* (these contain polar groups and O–H or N–H bonds), such as methanol (MeOH), the solvent can form hydrogen bonds to the anion. This lowers the nucleophilicity because a solvent shell surrounds the anion, and this hinders attack on the electrophile. Large anions are less solvated and hence are stronger nucleophiles than small anions in protic solvents. For example, the smaller F$^-$ ion is more heavily solvated than I$^-$ and hence is a weaker nucleophile.

4.2.1.2 Electrophiles

An electrophile with a positive charge is always a more powerful electrophile than its conjugate base (which is neutral).

Electrophilic strength

$$\overset{\overset{\oplus}{O}H}{\underset{C}{\|}} \longleftrightarrow \overset{OH}{\underset{\overset{\oplus}{C}}{|}} \quad > \quad \overset{O^{\delta-}}{\underset{\overset{\delta+}{C}}{\|}}$$

The relative electrophilic strength (or electrophilicity) of a cation depends on the stability of the positive charge. Inductive (+I), mesomeric (+M) and/or steric effects (see Section 4.4) can all lower the reactivity of the cation.

The relative electrophilic strength of an electrophilic site within a neutral molecule depends on the size of the partial positive charge ($\delta+$). Hydrogen or carbon atoms are electrophilic when attached to electronegative atoms (−I groups). The more electronegative the atom(s), the more electrophilic the hydrogen or carbon atom.

Electrophilic strength

$$\overset{\delta+}{H} \text{—} \overset{\delta-}{OH} \quad > \quad \overset{\delta+}{H} \text{—} \overset{\delta-}{NH_2}$$

The hydrogen atom in water is more electrophilic as oxygen is more electronegative than nitrogen

Electrophilic strength

$$\underset{R}{\overset{O^{\delta-}}{\underset{\diagdown}{\overset{\|}{\underset{Cl}{C^{\delta+}}}}}} \quad > \quad \underset{R}{\overset{O^{\delta-}}{\underset{\diagdown}{\overset{\|}{\underset{R}{C^{\delta+}}}}}}$$

acid chloride   *ketone*

The (carbonyl) carbon atom of the acid chloride is more electrophilic because both oxygen and chlorine atoms are electronegative

4.3 Carbocations, carbanions and carbon radicals

Carbocations, which include *carbenium* and *carbonium* ions, contain a positive charge on carbon. Carbenium ions have three bonds to the positively charged carbon (e.g. Me_3C^+), while carbonium ions contain five bonds (e.g. H_5C^+). (Carbenium ions are the most important.)

- *Carbenium ions* (R_3C^+) are generally planar and contain an empty p-orbital. They are stabilised by electron-donating groups (R = +I, +M) which delocalise the positive charge; +M groups are generally more effective than +I groups.

planar (sp^2)   planar (sp^2)   pyramidal (sp^3)

carbocation   carbanion

- *Carbanions* have three bonds on the carbon atom, which bears the negative charge. Carbanions (R_3C^-) can be planar (sp^2) or pyramidal

(sp^3) (or something in between). They are stabilised by: (i) electron-withdrawing groups (R = $-$I, $-$M); (ii) an increase in 's' character of the carbon bearing the negative charge; and (iii) aromatisation, which delocalises the negative charge (see Section 1.7).

- *Carbon radicals* have three bonds on the carbon atom, which contains the unpaired electron. Carbon radicals ($R_3C^{\bullet}$) are generally planar (sp^2). Like carbenium ions, they are stabilised by electron-donating groups (R = +I, +M) which delocalise the unpaired electron; +M groups are generally more effective than +I groups.

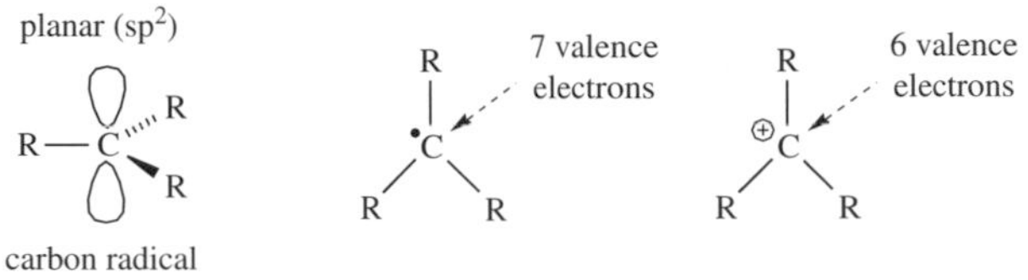

planar (sp^2)

carbon radical

Carbon radicals are similar to carbenium ions because both contain an electron-deficient carbon atom. For carbenium ions, the carbon atom is deficient of two electrons, and for carbon radicals, the carbon atom is deficient of one electron.

4.3.1 Order of stability

The order of stability of carbocations, carbanions and radicals bearing electron-donating (+I) alkyl groups is as follows.

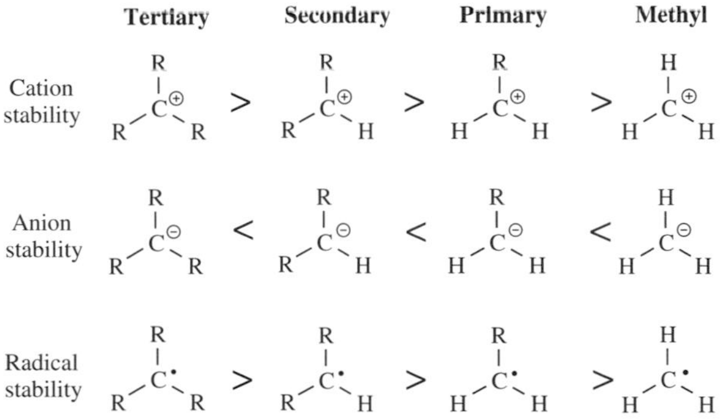

Carbanions can be stabilised by electron-withdrawing groups ($-$I, $-$M), whereas carbocations can be stabilised by electron-donating groups (+I, +M).

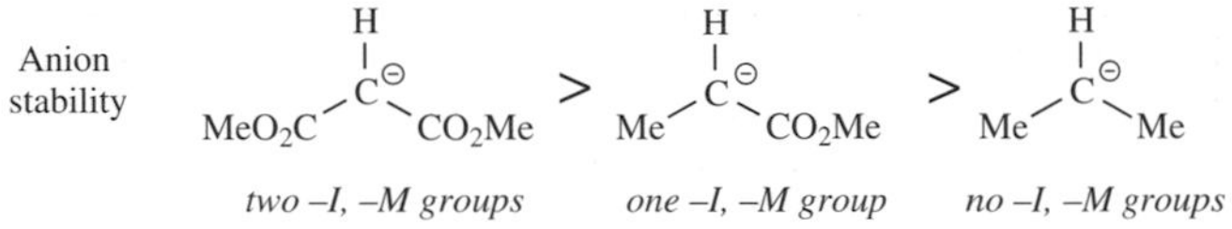

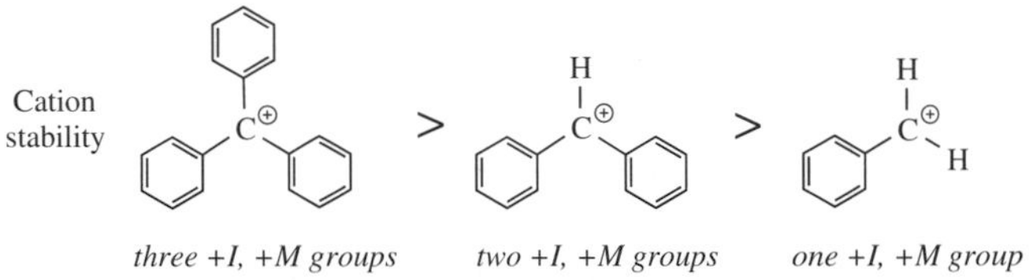

Anion
stability

$$\underset{\text{two }-I,-M\text{ groups}}{\overset{\overset{\displaystyle H}{\underset{\displaystyle |}{}}}{\underset{MeO_2C}{\overset{\ominus}{\underset{}{C}}}\diagdown CO_2Me}} \quad > \quad \underset{\text{one }-I,-M\text{ group}}{\overset{\overset{\displaystyle H}{\underset{\displaystyle |}{}}}{\underset{Me}{\overset{\ominus}{\underset{}{C}}}\diagdown CO_2Me}} \quad > \quad \underset{\text{no }-I,-M\text{ groups}}{\overset{\overset{\displaystyle H}{\underset{\displaystyle |}{}}}{\underset{Me}{\overset{\ominus}{\underset{}{C}}}\diagdown Me}}$$

Cation
stability

three +I, +M groups > *two +I, +M groups* > *one +I, +M group*

4.4   Steric effects

The size as well as the electronic properties (i.e. inductive and mesomeric effects) of the surrounding groups affects the stability of carbocations, carbanions and radicals. When bulky substituents surround a cation, for example, this reduces the reactivity of the cation to nucleophilic attack by *steric effects*. This is because the bulky groups hinder the approach of a nucleophile.

When the size of groups is responsible for reducing the reactivity at a site within a molecule, this is attributed to *steric hindrance*. When the size of groups is responsible for increasing the reactivity at a site within a molecule, this is attributed to *steric acceleration*.

Electronic and/or steric effects can explain the particular pathway for any given reaction.

4.5   Oxidation levels

- An *organic oxidation reaction* involves either (i) a decrease in the hydrogen content or (ii) an increase in the oxygen, nitrogen or halogen content of a molecule. Electrons are lost in oxidation reactions.
- An *organic reduction reaction* involves either (i) an increase in the hydrogen content or (ii) a decrease in the oxygen, nitrogen or halogen content of a molecule. Electrons are gained in reductions.

We can classify different functional groups by the *oxidation level* of the carbon atom within the functional group, using the following guidelines.

(1)   The greater the number of heteroatoms (e.g. O, N, halogen) attached to the carbon, the higher the oxidation level. Each bond to a heteroatom increases the oxidation level by +1.
(2)   The higher the degree of multiple bonding, the higher the oxidation level of the carbon.

Oxidation level	0	1	2	3	4
Functional groups	R_4C alkane	$RHC = CHR$ alkene	$RC \equiv CR$ alkyne		
		RCH_2X monohalide	$RCHX_2$ dihalide	RCX_3 trihalide	CX_4 tetrahalide
		RCH_2OR ether	$RCH(OR)_2$ acetal		
		RCH_2OH primary alcohol	$R-\overset{\displaystyle O}{\overset{\|}{C}}-R$ ketone	$R-\overset{\displaystyle O}{\overset{\|}{C}}-OH$ carboxylic acid	
		RCH_2NH_2 primary amine	$RCH = NH$ imine	$RC \equiv N$ nitrile	

- Carbon atoms (within functional groups) which are at the same oxidation level can be interconverted *without* oxidation or reduction.

Example (the numbers on each carbon are given in italic to indicate oxidation levels)

$$RC \equiv N \longrightarrow R-\overset{O}{\overset{\|}{C}}-OH \longrightarrow R-\overset{O}{\overset{\|}{C}}-OR \qquad \begin{array}{l}\text{No oxidation}\\\text{or reduction}\\\text{required}\end{array}$$
3   *3*   *3*

- Carbon atoms (within functional groups) which are at different oxidation levels are interconverted *by* oxidation (to increase the oxidation level) or reduction (to decrease the oxidation level).

Example (the numbers on each carbon are given in italic to indicate oxidation levels)

$$\xrightarrow{\hspace{2cm}} \text{oxidation}$$

$$RH_2C - CH_2R \longrightarrow RHC = CHR \longrightarrow RC \equiv CR$$
0   *0*   *1*   *1*   *2*   *2*

$$RCH_2OH \longrightarrow RCHO \longrightarrow RCO_2H$$
1   *2*   *3*

$$\xleftarrow{\hspace{2cm}} \text{reduction}$$

4.6   General types of reaction

4.6.1   Polar reactions (involving ionic intermediates)

4.6.1.1   Addition reactions

Addition reactions occur when two starting materials add together to form only one product.

$$A + B \longrightarrow A—B$$

The mechanism of these reactions can involve an initial electrophilic or nucleophilic attack on the key functional group.

Electrophilic addition to alkenes (see Section 6.2.2)

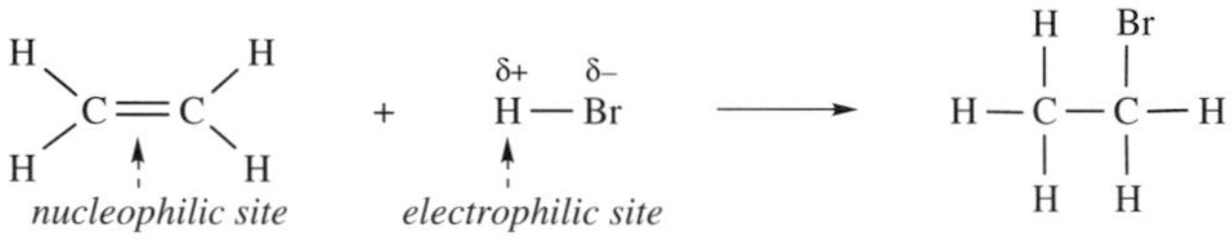

Nucleophilic addition to aldehydes and ketones (see Section 8.3)

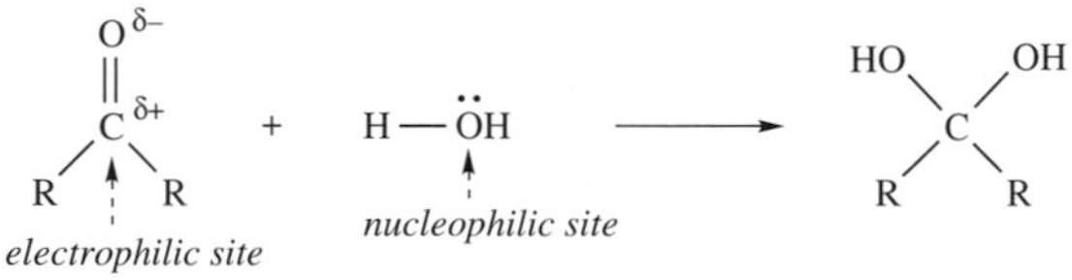

4.6.1.2   Elimination reactions

Elimination reactions are the opposite of addition reactions. One starting material is converted into two products.

$$A - B \longrightarrow A + B$$

The mechanism of these reactions can involve a loss of a cation or anion to form ionic reaction intermediates.

Elimination of alkyl halides (see Section 5.3.2)

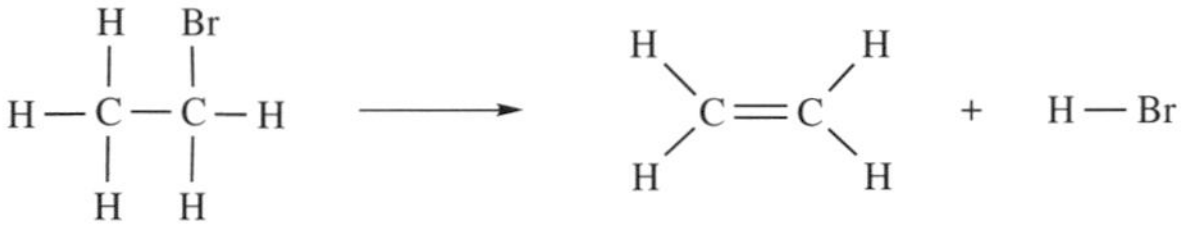

4.6.1.3   Substitution reactions

Substitution reactions occur when two starting materials exchange groups to form two new products.

$$A - B + C - D \longrightarrow A - C + B - D$$

The mechanism of these reactions can involve an initial electrophilic or nucleophilic attack on the key functional group.

Nucleophilic substitution of alkyl halides (see Section 5.3.1)

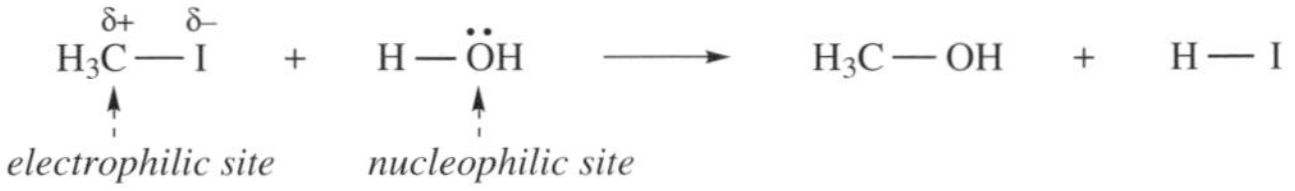

$$\underset{\substack{\uparrow \\ \textit{electrophilic site}}}{\overset{\overset{\delta+\ \ \delta-}{}}{H_3C-I}} \quad + \quad \underset{\substack{\uparrow \\ \textit{nucleophilic site}}}{H-\overset{..}{O}H} \quad \longrightarrow \quad H_3C-OH \quad + \quad H-I$$

Electrophilic substitution of benzene (see Section 7.2)

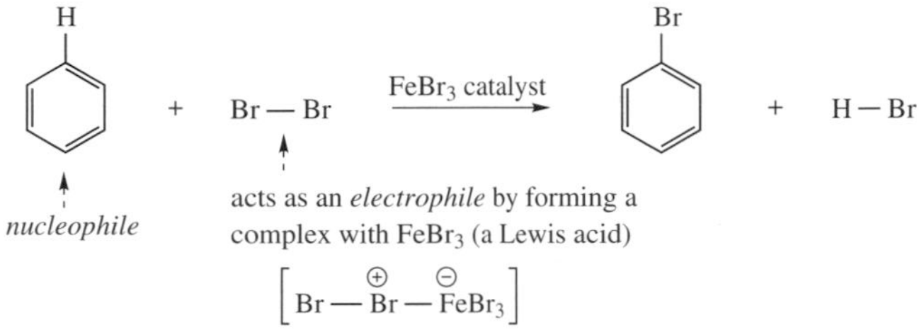

$$\left[Br \overset{\oplus}{-} Br \overset{\ominus}{-} FeBr_3 \right]$$

4.6.1.4 Rearrangement reactions

Rearrangement reactions occur when one starting material forms one product with a different arrangement of atoms and bonds (i.e. the product is an isomer of the starting material).

$$A \quad \longrightarrow \quad B$$

The mechanism of these reactions often involves carbocation intermediates, and the first-formed cation (e.g. primary or secondary) can rearrange to a more stable cation (e.g. tertiary).

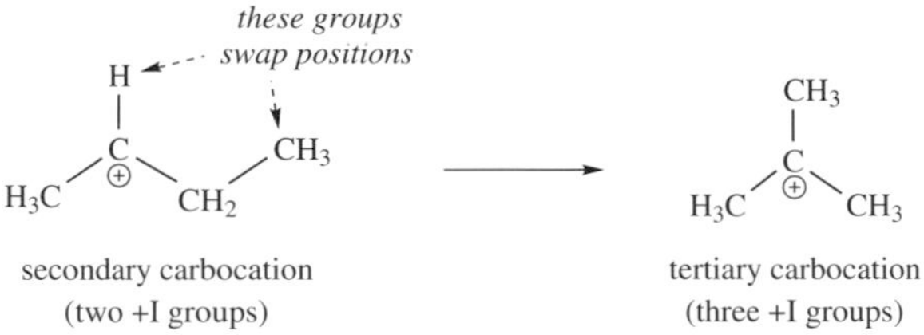

secondary carbocation
(two +I groups)

tertiary carbocation
(three +I groups)

4.6.2 Radical reactions

Radicals, like ions, can undergo addition, elimination, substitution and rearrangement reactions. A radical reaction comprises a number of steps.

(1) *Initiation* – the formation of radicals by homolytic bond cleavage (this generally requires heat or light).

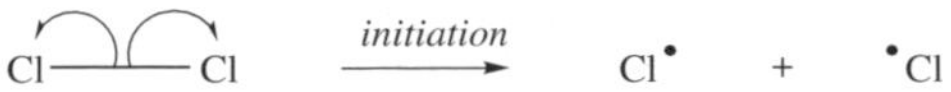

(2) *Propagation* – the reaction of a radical to produce a new product
     radical. This may involve an addition, elimination, substitution or
     rearrangement reaction.

propagation reactions

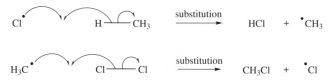

(3) *Termination* – the coupling of two radicals to form only non-radical
     products.

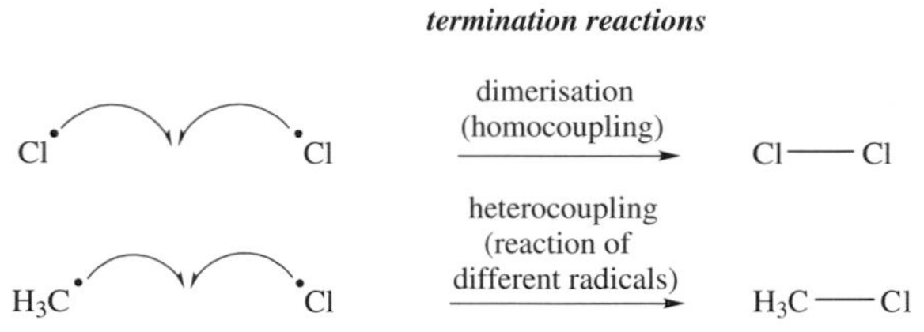

termination reactions

The chlorination of alkanes, such as methane, leads to the substitution
of hydrogen atoms for chlorine atoms in a radical chain reaction (see
Section 5.2.1).

4.6.3 Pericyclic reactions

Pericyclic reactions take place in a single step without (ionic or
radical) intermediates and involve a cyclic redistribution of bonding
electrons.

The Diels–Alder cycloaddition reaction (see Section 6.2.2.11)

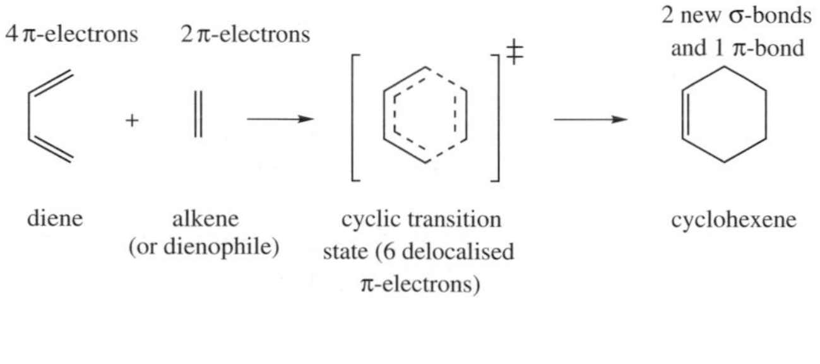

4π-electrons 2π-electrons   2 new σ-bonds
                                                                 and 1 π-bond

diene   alkene   cyclic transition   cyclohexene
               (or dienophile) state (6 delocalised
                                π-electrons)

4.7 Ions versus radicals

- *Heterolytic cleavage* of bonds occurs at room temperature in polar solvents. The ions which are formed are solvated (i.e. a solvent shell surrounds them) and stabilised by polar solvents.
- *Homolytic cleavage* of bonds occurs at high temperature in the absence of polar solvents. When a compound is heated in a non-polar solvent, radicals are formed. Radicals are uncharged and hence have little interaction with the solvent. The energy required to cleave a bond homolytically, to give radicals, is called the *bond dissociation energy* or the *bond strength*. The lower the bond dissociation energy, the more stable the radicals (as they are easier to form).
- *Ionic reactions* occur because of electrostatic attraction; a positive or δ+ charge attracts a negative or δ– charge. Electron-rich sites react with electron-deficient sites.
- *Radical reactions* occur because radicals, which have an odd number of electrons in the outer shell, need to 'pair' the electron to produce a filled outer shell.

4.8 Reaction selectivity

- *Chemoselectivity* – reaction at one functional group in preference to another functional group(s).

Example: reduction of a ketoester (see Section 8.3.3.1)

ketone   ester   secondary
                                                   alcohol   ester

O   O   NaBH₄   OH   O

          OMe   ─────────▶   OMe
                     then H₂O

selective reduction of the ketone to the secondary alcohol

- *Regioselectivity* – reaction at one position within a molecule in preference to others. This leads to the selective formation of one regioisomer.

Example: addition of HCl to an unsymmetrical alkene (see Section 6.2.2.1)

                                                 Cl   H
Me | |
   C=CH₂   ─── H–Cl ───▶   Me — C — CH₂
Me |
                                                 Me

regioselective addition of the H atom onto the carbon at the end of the double
bond and the Cl atom onto the carbon with the most alkyl substituents

- *Stereoselectivity* – the formation of one enantiomer, one diastereoisomer or one double bond isomer in preference to others.

Example: catalytic hydrogenation of an alkyne (see Section 6.3.2.4)

$$\text{Et}-\text{C}\equiv\text{C}-\text{Et} \quad \xrightarrow[\substack{\text{Lindlar catalyst} \\ \text{(Pd/CaCO}_3\text{/PbO)}}]{\text{H}_2} \quad \substack{\text{Et} \\ \diagdown \\ \text{H}} \text{C}=\text{C} \substack{\diagup \text{Et} \\ \diagdown \text{H}}$$

cis- or Z-alkene

formation of a Z- rather than an E-alkene

4.9 Reaction thermodynamics and kinetics

The *thermodynamics* of a reaction tells us in what direction the reaction proceeds (and how much energy will be consumed or released). The *kinetics* of a reaction tells us whether the reaction is fast or slow.

4.9.1 Thermodynamics

4.9.1.1 Equilibria

All chemical reactions can be written as equilibrium processes, in which forward and backward reactions occur concurrently to give an equilibrium position. Drawing arrows pointing in the opposite directions denotes the equilibrium, and the position of the equilibrium is expressed by the *equilibrium constant* (K). For simplicity, this is defined below in terms of concentration rather than activity (as activity is approximately equal to concentration in dilute solution).

$$\text{A} \rightleftharpoons \text{B} \qquad K = \frac{\text{concentration of product(s) at equilibrium}}{\text{concentration of reactant(s) at equilibrium}} = \frac{[\text{B}]_{\text{eq}}}{[\text{A}]_{\text{eq}}}$$

If K is larger than 1, then the concentration of B will be larger than the concentration of A at equilibrium.

The size of K is related to the free energy difference between the starting materials and products. For the efficient conversion of A to B, a high value of K is required. This means that the free energy of the products (in kJ mol^{-1}) must be lower than the free energy of the starting materials. The total free energy change (under standard conditions) during a reaction is called the *standard Gibbs free energy change* – ΔG° (kJ mol^{-1}).

$$\Delta G^{\circ} \;=\; \text{free energy of products } - \text{ free energy of reactants}$$

$$\Delta G^{\circ} \;=\; -RT \ln K \qquad \begin{array}{l} R = \text{gas constant (8.314 J K}^{-1}\text{ mol}^{-1}\text{)} \\ T = \text{absolute temperature (in K)} \end{array}$$

The standard free energy change for a reaction is a reliable guide to the extent of a reaction, provided that equilibrium is reached.

- if $\Delta G°$ is *negative*, then the *products*, will be favoured at equilibrium ($K > 1$);
- if $\Delta G°$ is *positive*, then the *reactants* will be favoured at equilibrium ($K < 1$);
- if $\Delta G°$ is *zero*, then $K = 1$, and hence there will be the same concentration of reactants and products.

At a particular temperature, K is constant.

Remember that K and $\Delta G°$ say nothing about the rate of a reaction. The sign and magnitude of $\Delta G°$ decide whether an equilibrium lies in one direction or another.

4.9.1.2 Enthalpy and entropy

The standard Gibbs free energy change of reaction, $\Delta G°$ is related to the *enthalpy change of reaction* ($\Delta H°$) and *the entropy change of reaction* ($\Delta S°$). The change in free energy for a reaction at a given temperature has contributions from *both* the change in enthalpy and the change in entropy.

$$\Delta G° = \Delta H° - T\Delta S°$$
T = absolute temperature (in K)

Enthalpy

The enthalpy change of reaction is the heat exchanged with the surroundings (at constant pressure) in a chemical reaction. This represents the difference in stability (bond strength) of the reagents and products.

- if $\Delta H°$ is negative, then the bonds in the product(s) are *stronger* overall than those in the starting material. Heat is released in an *exothermic* reaction;
- if $\Delta H°$ is positive, then the bonds in the product(s) are *weaker* overall than those in the starting material. Heat is absorbed in an *endothermic* reaction.

Entropy

The entropy change of reaction provides a measure of the change in molecular disorder or randomness caused by a reaction.

- $\Delta S°$ is negative when the reaction leads to *less* disorder. This occurs when two reactants are converted to one product.
- $\Delta S°$ is positive when the reaction leads to *more* disorder. This occurs when one reactant is converted to two products.

Gibbs free energy

For a negative value of $\Delta G°$ (in which products are favoured over reactants at equilibrium), we require low positive, or high negative, values of $\Delta H°$ and high positive values of $T\Delta S°$.

- if the reaction is *exothermic*, then $\Delta H°$ will be *negative*, and hence $\Delta G°$ will be negative, provided $\Delta S°$ is not large and negative;
- if the reaction is *endothermic*, then $\Delta H°$ will be *positive* and the $-T\Delta S°$ term will need to be larger than this for $\Delta G°$ to be negative. This will require a large, positive entropy change ($\Delta S°$) and/or a high reaction temperature (T).

4.9.2 Kinetics

4.9.2.1 Reaction rate

Although a negative value of $\Delta G°$ is required for a reaction to occur, the rate at which it occurs is determined by the *activation energy*, $\Delta G^{\ddagger}$ or E_a. This is the energy difference between the reactants and the *transition state*. A transition state is a structure that represents an energy maximum on converting starting materials to products, and it cannot be isolated. It is often drawn in square brackets with a superscript double-dagger ($\ddagger$). (This is not the same as a reaction *intermediate*, which occurs at a local energy minimum and can be detected and sometimes isolated (see below).)

We can show the energy changes that occur during a reaction by an energy diagram.

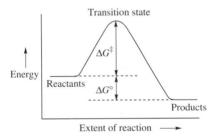

The higher the activation energy, the slower the reaction. Reactions having activation energies above $20\,kJ\,mol^{-1}$ generally require heating, so that reactants can be converted into products. An increase in temperature therefore increases the rate of the reaction. Most organic reactions have activation energies of between 40 and $150\,kJ\,mol^{-1}$.

The activation energy can be determined experimentally by measuring the *rate constant* of the reaction at different temperatures using the *Arrhenius equation*.

$k = Ae^{-E_a/RT}$	k = rate constant	R = gas constant (J K^{-1} mol^{-1})
	A and e = constants	T = temperature (K)

- The rate constant, k, can be determined by monitoring the rate at which starting materials disappear/products appear (at constant temperature), by varying the concentration of the reactant or reactants.

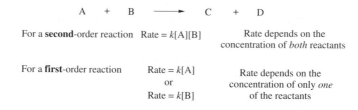

For a **second**-order reaction   Rate = k[A][B]   Rate depends on the
                                                                    concentration of *both* reactants

For a **first**-order reaction   Rate = k[A]   Rate depends on the
                                              or   concentration of only *one*
                                      Rate = k[B]   of the reactants

Organic reactions generally consist of a number of successive steps. The slowest step (which leads to the highest energy transition state) is called the *rate-determining step*, and this is the reaction rate that can be determined experimentally.

Organic reactions with a number of steps will have *intermediates*. These represent a localised minimum energy in the reaction profile. An energy barrier must be overcome before the intermediate forms a more stable product(s) or a second intermediate.

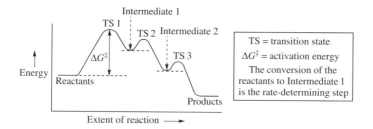

The structure of the intermediates can give us an idea of the structure of the transition states. The *Hammond postulate* states that the structure of a transition state resembles the structure of the nearest stable species.

- For an *exothermic* reaction (or step of a reaction), the transition state resembles the structure of the *reactant*. This is because the energy level of the transition state is closer to the reactant, than to the product.

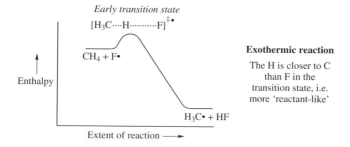

- For an *endothermic* reaction (or step of a reaction), the transition state resembles the structure of the *product*. This is because the energy level of the transition state is closer to the product, than to the reactant.

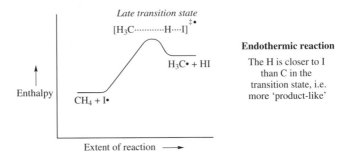

Late transition state

$[H_3C\cdots\cdots H\cdots I]^{\ddagger\bullet}$

Endothermic reaction

$H_3C\bullet + HI$

The H is closer to I than C in the transition state, i.e. more 'product-like'

Enthalpy

$CH_4 + I\bullet$

Extent of reaction ⟶

Catalysts increase the rate of a reaction by allowing the reaction to proceed by a different pathway, which has a lower energy transition state. Although the catalyst affects the rate at which an equilibrium is established, it does not alter the position of the equilibrium. The addition of a catalyst, which is not chemically changed during the reaction, allows many reactions to take place more quickly and at lower temperatures.

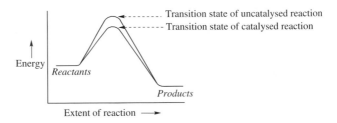

- - - - - - - Transition state of uncatalysed reaction
- - - - - - - Transition state of catalysed reaction

Energy

Reactants

Products

Extent of reaction ⟶

A *homogeneous* catalyst is in the *same* phase as the starting materials of the reaction that it is catalysing.

A *heterogeneous* catalyst is in a *different* phase from the starting materials of the reaction that it is catalysing.

4.9.3 Kinetic versus thermodynamic control

For a reaction that can give rise to more than one product, the amount of each of the different products can depend on the reaction temperature. This is because, although all reactions are reversible, it can be difficult to reach equilibrium, and a non-equilibrium ratio of products can be obtained.

- At *low* temperatures, reactions are more likely to be *irreversible* and equilibrium is less likely to be reached. Under these conditions, the product that is formed at the fastest rate predominates. This is *kinetic control*. The *kinetic product* is therefore formed at the fastest rate (i.e. this product has the lowest activation energy barrier).
- At *high* temperatures, reactions are more likely to be *reversible* and an equilibrium is likely to be reached. Under these conditions, the energetically more stable product predominates. This is *thermodynamic control*. The *thermodynamic product* is therefore the most stable product (i.e. this product has the lowest energy).

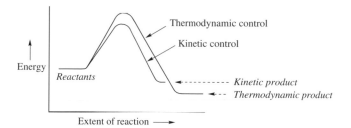

4.10   Orbital overlap and energy

Two atomic orbitals can combine to give two molecular orbitals – one *bonding* molecular orbital (lower in energy than the atomic orbitals) and one *antibonding* molecular orbital (higher in energy than the atomic orbitals) (see Section 1.4). Orbitals that combine in-phase form a bonding molecular orbital, and for the best orbital overlap, the orbitals should be of the same size.

The orbitals can overlap end-on (as for σ-bonds) or side-on (as for π-bonds). The empty orbital of an electrophile (which accepts electrons) and the filled orbital of a nucleophile (which donates electrons) will point in certain directions in space. For the two to react, the filled and empty orbital must be correctly aligned; for end-on overlap, the filled orbital should point directly at the empty orbital.

The orbitals must also have a similar energy. For the greatest inter-action, the two orbitals should have the same energy. Only the *highest energy occupied orbitals* (or HOMOs) of the nucleophile are likely to be similar in energy to the *lowest energy unoccupied orbitals* (or LUMOs) of the electrophile.

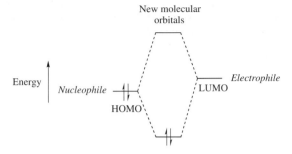

The two electrons enter a lower energy molecular orbital. There is
therefore a gain in energy and a new bond is formed. The further
apart the HOMO and LUMO, the lower the gain in energy

- The HOMO of a nucleophile is usually a (non-bonding) lone pair
  or a (bonding) π-orbital. (These are higher in energy than a σ-orbital.)
- The LUMO of an electrophile is usually a (antibonding) π*-orbital.
  (This is lower in energy than a σ*-orbital.)

Example: addition of water to a ketone

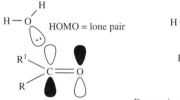

For maximum orbital overlap attack at 90° is
required. However, attack at ~107° is observed
because of the greater electron density on the
carbonyl oxygen atom, which repels the lone
pair on H_2O

4.11 Guidelines for drawing reaction mechanisms

(1) Draw the structure of the reactants, and include lone pairs on any
    heteroatoms. Show bond polarities by δ+ and δ–.
(2) Choose which reactant is the nucleophile and which is the elec-
    trophile, and identify the nucleophilic atom in the nucleophile
    and the electrophilic atom in the electrophile.
(3) Draw a curly arrow from the nucleophilic atom to the electrophilic
    atom to make a new bond. The arrow can start from a negative
    charge, a lone pair of electrons or a multiple bond. If a new bond
    is made to an uncharged H, C, N or O atom in the electrophile,

then one of the existing bonds should be broken and a second curly arrow should be drawn (pointing in the same direction as the first).

(4) The overall charge of the reactants should be the same as that of the products. Try to end up with a negative charge on an electronegative atom.

- If the *nucleophilic* atom is neutral, then the atom will gain a positive charge. If it is negatively charged, then it will become neutral.
- If the *electrophilic* atom is neutral, then one of the existing bonds must be broken, and a negative charge will reside on the most electronegative atom. If it is positively charged, then it will become neutral.

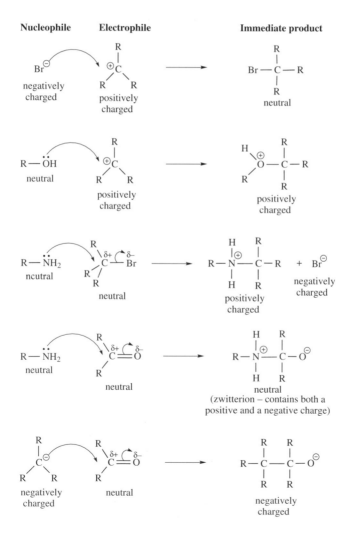

Problems

(1) Use oxidation levels to help deduce if the following reactions
     involve reduction or oxidation? If so, state if the reaction involves
     oxidation or reduction of the organic starting materials.

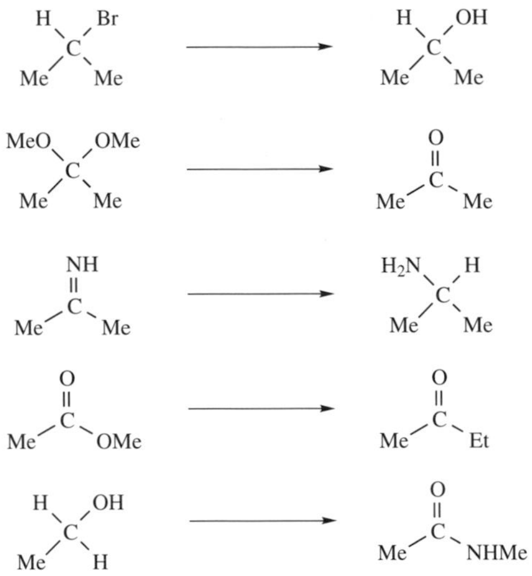

(2) For the following reactions, state which starting material can act
     as the nucleophile and which can act as the electrophile. Draw the
     likely product from each of the reactions and show its formation
     by using curly arrows.

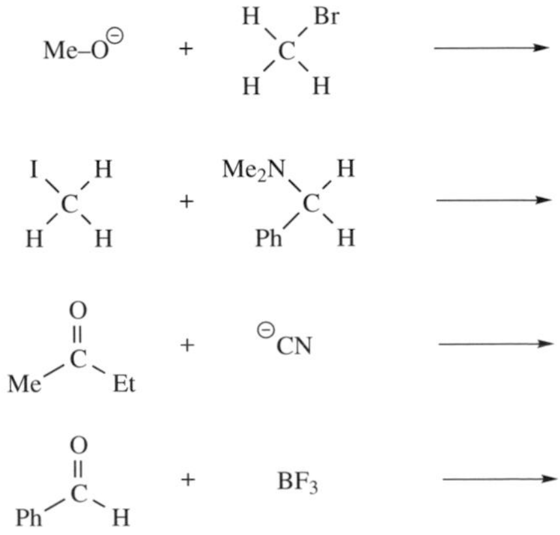

(3) Classify the following reactions as addition, elimination, substitution or rearrangement reactions.

(a) 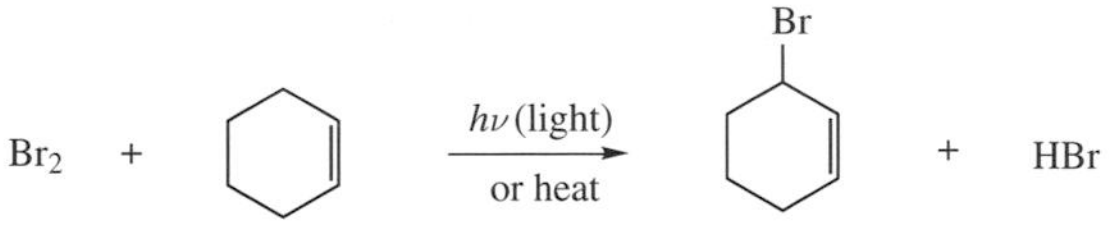 Br_2 + $\xrightarrow[\text{or heat}]{h\nu\,(\text{light})}$ + HBr

(b) 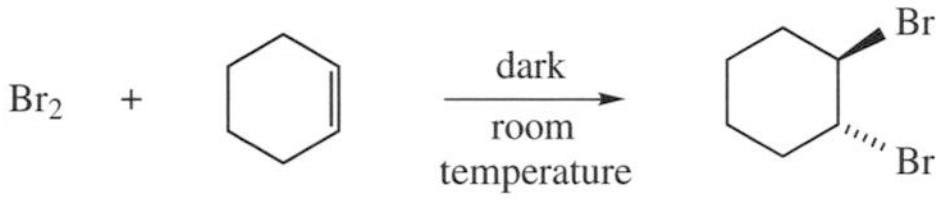 Br_2 + $\xrightarrow[\substack{\text{room} \\ \text{temperature}}]{\text{dark}}$

(c) 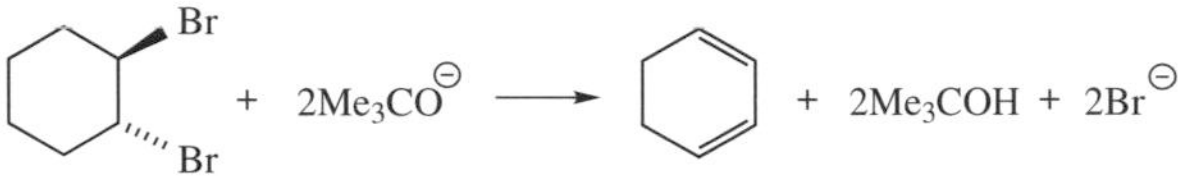 + $2Me_3CO^{\ominus}$ ⟶ + $2Me_3COH$ + $2Br^{\ominus}$

(d) 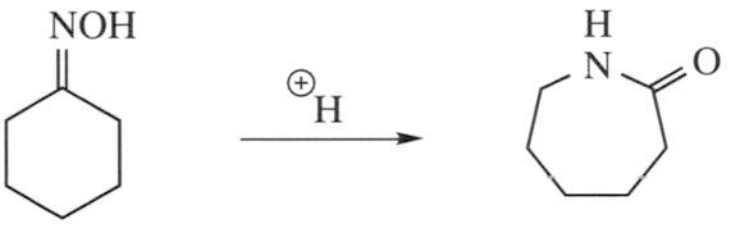 $\xrightarrow{\overset{\oplus}{H}}$

(4) Classify the following reactions as chemoselective, regioselective or stereoselective reactions.

(a) 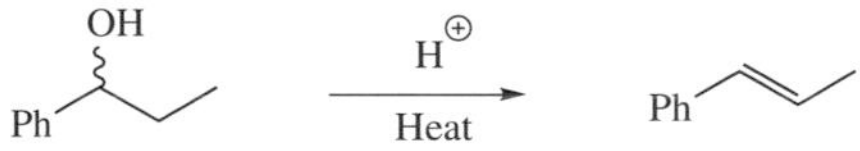 $\xrightarrow[\text{Heat}]{\overset{\oplus}{H}}$

(b)  $\xrightarrow{Me_3CO^{\ominus}}$

(c) 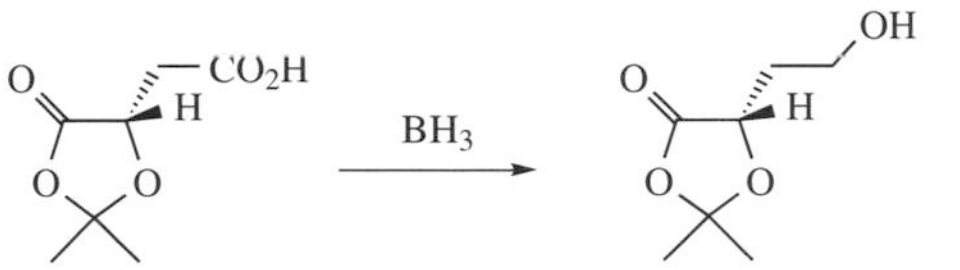 $\xrightarrow{BH_3}$

(d) 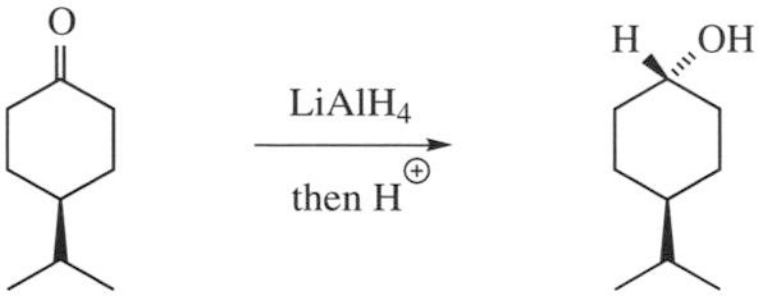 $\xrightarrow[\text{then } \overset{\oplus}{H}]{LiAlH_4}$

5. ALKYL HALIDES

Key point. Alkyl halides are composed of an alkyl group bonded to a halogen atom (X = F, Cl, Br, I). As halogen atoms are more electronegative than carbon, the C−X bond is polar and nucleophiles can attack the slightly positive carbon atom. This leads to the halogen atom being replaced by the nucleophile in a *nucleophilic substitution reaction*, and this can occur by either an S_N1 (*two-step*) *mechanism* or an S_N2 (*concerted* or *one-step*) *mechanism*. In competition with substitution is *elimination*, which results in the loss of HX from alkyl halides to form alkenes. This can occur by either an E1 (*two-step*) *mechanism* or an E2 (*concerted*) *mechanism*. The mechanism of the substitution or elimination reaction depends on the alkyl halide, the solvent and the nucleophile/base.

5.1 Structure

Alkyl halides have an alkyl group joined to a halogen atom by a single bond. The larger the size of the halogen atom (X = I > Br > Cl), the weaker the C−X bond (Appendix 1). The C−X bond is polar, and the carbon atom bears a slight positive charge and the electronegative halogen atom bears a slight negative charge.

$$\delta+ \quad \delta- \qquad R = \text{alkyl or H}$$
$$R_3C \text{---} X \qquad X = F, Cl, Br, I$$

sp^3 hybridisation (tetrahedral)

Alkyl halides have an sp^3 carbon atom and hence have a tetrahedral shape.

- An *aliphatic* alkyl halide has the carbon atoms in a chain and not a closed ring, e.g. 1-bromohexane, $CH_3(CH_2)_5Br$.
- An *alicyclic* alkyl halide has the carbon atoms in a closed ring, but the ring is not aromatic, e.g. bromocyclohexane, $C_6H_{11}Br$.
- An *aromatic* alkyl halide has the carbon atoms in a closed ring, and the ring is aromatic, e.g. bromobenzene, C_6H_5Br.

5.2 Preparation

5.2.1 Halogenation of alkanes

Alkyl chlorides or bromides can be obtained from alkanes by reaction with chlorine or bromine gas, respectively, in the presence of UV light. The reaction involves a radical chain mechanism.

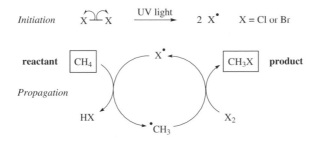

Initiation $X \overset{\frown}{\underset{\smile}{X}}$ $\xrightarrow{\text{UV light}}$ $2 \; X^{\bullet}$ $\quad X = Cl \; or \; Br$

reactant $\boxed{CH_4}$ $\quad X^{\bullet}$ $\quad \boxed{CH_3X}$ **product**

Propagation

$HX \quad \quad \quad \quad X_2$

$^{\bullet}CH_3$

This is a *substitution* reaction, as a hydrogen atom on the carbon is substituted for a Cl or Br atom. A mixture of halogenated products is usually obtained if further substitution reactions can take place.

$$CH_4 \xrightarrow[]{X_2 \; HX} CH_3X \xrightarrow[]{X_2 \; HX} CH_2X_2 \xrightarrow[]{X_2 \; HX} CHX_3 \xrightarrow[]{X_2 \; HX} CX_4$$

The ease of halogenation depends on whether the hydrogen atom is bonded to a primary, secondary or tertiary carbon atom. A tertiary hydrogen atom is more reactive because reaction with a halogen atom ($X^{\bullet}$) produces an intermediate tertiary radical, which is more stable (and therefore more readily formed) than a secondary or primary radical (see Section 4.3).

Order of reactivity
toward $X^{\bullet}$ $\quad R_3C \!-\! H \quad > \quad R_2CH \!-\! H \quad > \quad RCH_2 \!-\! H$

Order of radical
stability $\quad R_3\overset{\bullet}{C} \quad > \quad R_2\overset{\bullet}{CH} \quad > \quad R\overset{\bullet}{C}H_2$

$\quad\quad\quad$ *tertiary* $\quad\quad\quad$ *secondary* $\quad\quad\quad$ *primary*

5.2.2 Halogenation of alcohols

Alcohols are converted to alkyl halides using a number of methods. All methods involve 'activating' the OH group to make this into a better leaving group (see Section 5.3.1.4).

$$X^{\ominus} \; + \; R \!-\! \overset{\frown}{OH} \; \xrightarrow{\; \times \;} \; X \!-\! R \; + \; {}^{\ominus}OH$$

The hydroxide ion is relatively unstable and hence is a bad leaving group

The mechanism of these reactions depends on whether a primary, secondary or tertiary alcohol is used. (Section 5.3.1 discusses the mechanisms.)

Reaction with HX (X = Cl, Br, I)

The OH group can be converted into a better leaving group, namely water, by protonation. As water is neutral, it is more stable than the hydroxide ion and hence is a better leaving group.

- *Primary alcohol.* This proceeds *via* nucleophilic attack on an alkyloxonium ion (S_N2 mechanism).

concerted mechanism

$$RCH_2\overset{..}{O}H + H \overset{\frown}{\smile} X \longrightarrow X^{\ominus}\ RCH_2 \overset{\oplus}{C}O\overset{H}{\underset{H}{}} \longrightarrow X{-}CH_2R + H_2O$$

alkyloxonium ion

- *Tertiary alcohol.* This proceeds *via* the loss of water from the alkyloxonium ion (to give a relatively stable tertiary carbocation) before nucleophilic attack can take place (S_N1 mechanism).

$$R_3C-\overset{..}{O}H + H\overset{\frown}{\smile}X \xrightarrow{step\ 1} R_3C\overset{\oplus}{O}\overset{H}{\underset{H}{}} + X^{\ominus}$$

$$step\ 2 \downarrow -H_2O$$

$$R_3C-X \longleftarrow R_3\overset{\oplus}{C} + X^{\ominus}$$

- *Secondary alcohol.* This can proceed *via* either an S_N1 mechanism or an S_N2 mechanism.

Reaction with phosphorus trihalides (PBr$_3$, PCl$_3$)

The OH group is converted into a neutral HOPX$_2$-leaving group (by an initial S_N2 reaction at phosphorus).

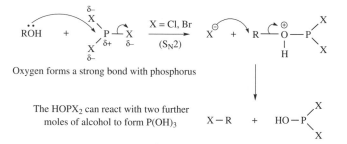

Oxygen forms a strong bond with phosphorus

The HOPX$_2$ can react with two further moles of alcohol to form P(OH)$_3$

$$X-R + HO-P\overset{X}{\underset{X}{}}$$

Reaction with thionyl chloride (SOCl$_2$)

Alkyl chlorides can be prepared by the reaction of alcohols with thionyl chloride (SOCl$_2$) in the presence of a nitrogen base (e.g. triethylamine or pyridine). An intermediate alkyl chlorosulfite (ROSOCl) is formed by nucleophilic attack of ROH on the sulfur atom of thionyl chloride. The OH group is converted to an OSOCl-leaving group, which is displaced on reaction with the chloride anion (e.g. in an S_N2 mechanism when R is a primary alkyl group).

The intermediate alkyl chlorosulfite could be formed via an intermediate dichloride

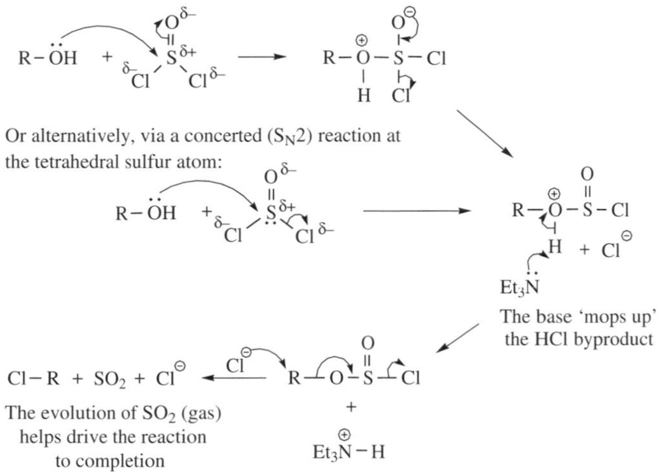

Or alternatively, via a concerted (S_N2) reaction at the tetrahedral sulfur atom:

The base 'mops up' the HCl byproduct

The evolution of SO_2 (gas) helps drive the reaction to completion

The mechanism changes to S_Ni in the absence of a nitrogen base (see Section 5.3.1.7).

Reaction with p-toluenesulfonyl chloride

Alkyl chlorides, bromides and iodides can be formed by the reaction of alcohols with p-toluenesulfonyl chloride (or tosyl chloride, abbreviated as TsCl) in the presence of a nitrogen base (e.g. triethylamine or pyridine). The OH group is converted into a tosylate (abbreviated as ROTs), which can be displaced on reaction with Cl⁻, Br⁻ or I⁻. The stable tosylate anion is an excellent leaving group (S_N1 or S_N2 mechanism depending on the nature of the alkyl group, R).

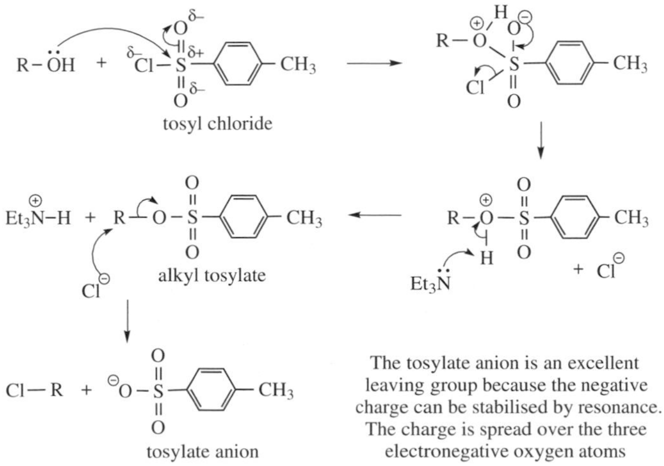

tosyl chloride

alkyl tosylate

tosylate anion

The tosylate anion is an excellent leaving group because the negative charge can be stabilised by resonance. The charge is spread over the three electronegative oxygen atoms

5.2.3   Halogenation of alkenes

The electrophilic addition of HX or X_2 to alkenes generates alkyl halides with one or two halogen atoms, respectively (see Section 6.2.2 for a detailed discussion of the mechanisms).

Electrophilic addition of Br$_2$

The electron-rich alkene double bond repels the electrons in the bromine molecule to create a partial positive charge on the bromine atom near the double bond. An intermediate bromonium ion is formed, which reacts to give the *trans*-dibromide derived from *anti*- addition (i.e. the two Br groups add to the alkene from the *opposite* sides).

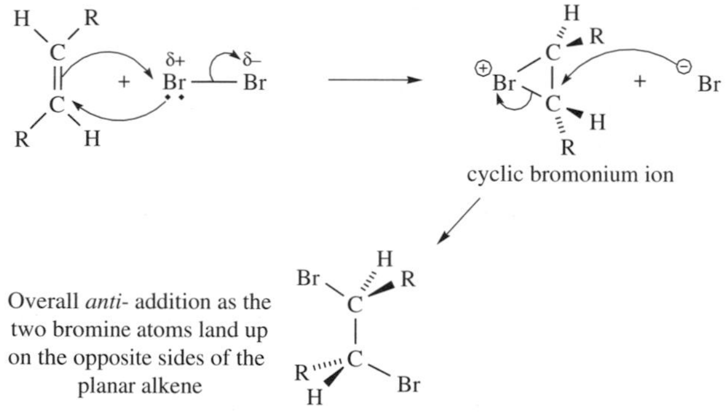

cyclic bromonium ion

Overall *anti*- addition as the
two bromine atoms land up
on the opposite sides of the
planar alkene

As the product with the *anti*- stereochemistry is formed in excess over the *syn*- addition product (in which the two Br groups add to the alkene from the *same* side), the reaction is *stereoselective* (i.e. one particular stereoisomer of the product is formed in excess).

Electrophilic addition of HX

The electron-rich alkene double bond reacts with a proton so as to make the most stable intermediate carbocation. The addition is regio-selective, and the so-called Markovnikov (also spelt Markownikoff) product is formed. If a peroxide is added, the reaction proceeds *via* radical intermediates to give the anti-Markovnikov product.

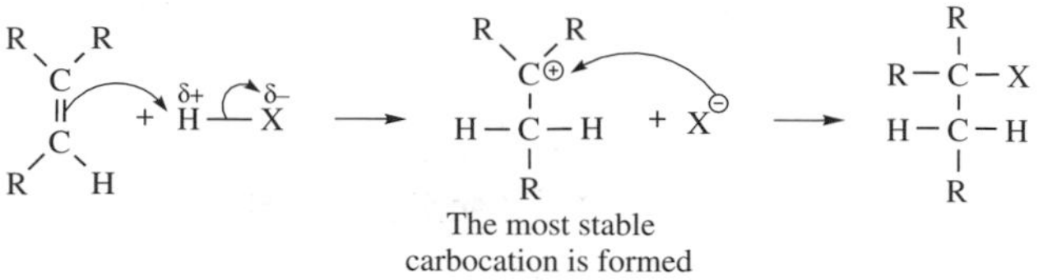

The most stable
carbocation is formed

5.3   Reactions

Alkyl halides react with nucleophiles in *substitution* reactions and with bases in *elimination* reactions.

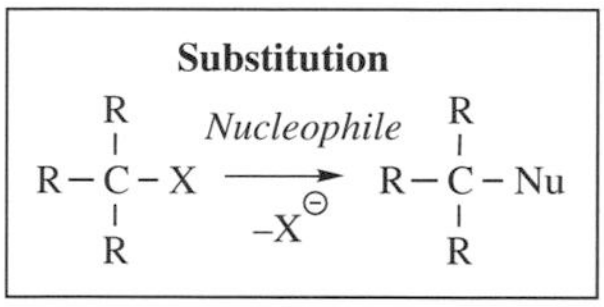

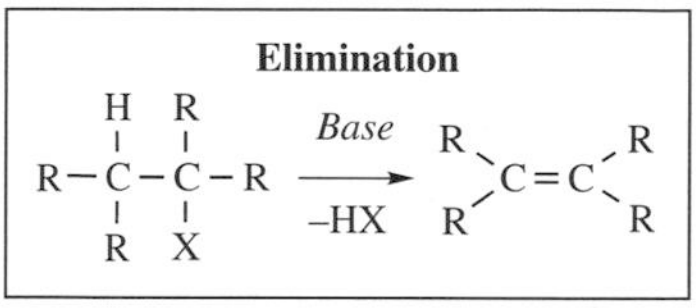

5.3.1   Nucleophilic substitution

The two main mechanisms for nucleophilic substitution of alkyl halides are S_N1 and S_N2. These represent the extreme mechanisms of nucleophilic substitution, and some reactions involve mechanisms which lie somewhere in between the two.

In both S_N1 and S_N2 reactions, the mechanisms involve the loss of the halide anion (X^-) from RX. The halide anion that is expelled is called the *leaving group*.

5.3.1.1   The S_N2 (substitution, nucleophilic, bimolecular) reaction

This is a concerted (one-step) mechanism in which the nucleophile forms a new bond to carbon at the same time as the bond to the halogen is broken. The reaction is second order or bimolecular, as the rate depends on the concentration of both the nucleophile and the alkyl halide.

$$\text{Reaction rate} = k[\text{Nu}][\text{RX}]$$

The reaction leads to an inversion (or change) of stereochemistry at a chiral centre (i.e. an *R*-enantiomer will be converted to an *S*-enantiomer). This is known as a *Walden inversion*.

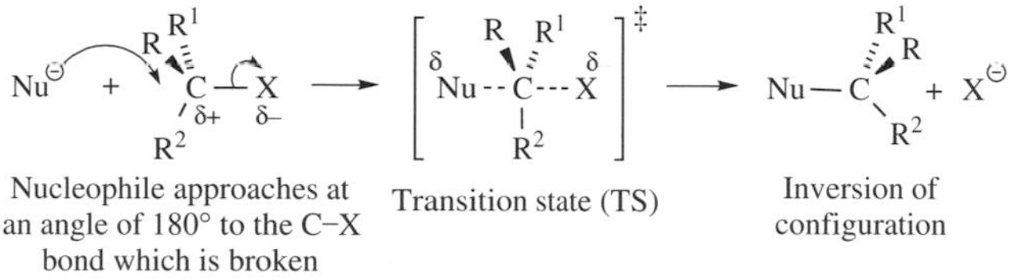

Nucleophile approaches at an angle of 180° to the C–X bond which is broken

Transition state (TS)

Inversion of configuration

The charge is spread from the nucleophile to the leaving group.
The carbon atom in the TS is partially bonded to Nu and X

The nucleophile approaches the C–X bond at an angle of 180°. This maximises the interaction of the filled orbital of the nucleophile with the empty σ^*-orbital of the C–X bond.

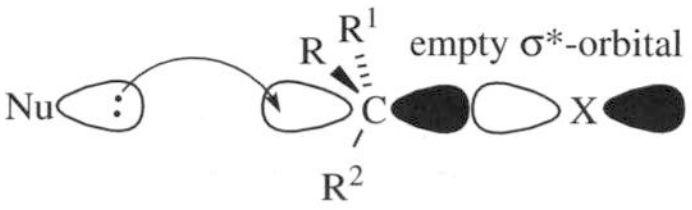

Alkyl halides with bulky alkyl substituents react more slowly than those with small alkyl substituents on the central carbon atom. Bulky substituents prevent the nucleophile from approaching the central carbon atom. S_N2 reactions can therefore only occur at relatively unhindered sites.

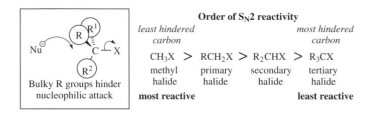

Order of S_N2 reactivity

least hindered carbon			most hindered carbon
CH_3X >	RCH_2X >	R_2CHX >	R_3CX
methyl halide	primary halide	secondary halide	tertiary halide
most reactive			**least reactive**

Bulky R groups hinder nucleophilic attack

The presence of many +I alkyl groups on the central carbon atom also reduces the partial positive charge, which reduces the rate at which the nucleophile attacks the carbon atom.

5.3.1.2   The S_N1 (substitution, nucleophilic, unimolecular) reaction

This is a stepwise mechanism involving initial cleavage of the carbon–halogen bond to form an intermediate carbocation. The reaction is first order, or unimolecular, as the rate depends on the concentration of only the alkyl halide (and *not* the nucleophile).

$$\text{Reaction rate} = k[\text{RX}]$$

The reaction leads to the racemisation of a stereogenic centre in the starting material (i.e. an *R*-enantiomer will be converted to a 50:50 mixture of *R*- and *S*-enantiomers). This is because the nucleophile can equally attack either side of the planar carbocation.

The more stable the carbocation intermediate, the faster the S_N1 reaction (i.e. the easier it is to break the C–X bond). Tertiary halides will therefore react faster than primary halides by this mechanism, because a tertiary carbocation is more stable than a primary carbocation.

Order of S$_N$1 reactivity

produces most stable carbocation			*produces least stable carbocation*
R$_3$CX	> R$_2$CHX	> RCH$_2$X	> CH$_3$X
tertiary halide	secondary halide	primary halide	methyl halide
most reactive			**least reactive**

Primary benzylic and allylic halides can undergo S$_N$1 reactions because the carbocations are stabilised by resonance. These carbocations have stability similar to that of secondary alkyl carbocations.

Benzylic halide

benzylic cation

Allylic halide

allylic cation

5.3.1.3 S$_N$2 reactions versus S$_N$1 reactions

The mechanism of a nucleophilic substitution reaction is influenced by the nature of the alkyl halide, the nucleophile and the solvent.

The alkyl halide

- *Primary* alkyl halides are likely to react by an S$_N$2 mechanism.
- *Secondary* alkyl halides are likely to react by either an S$_N$1 mechanism or an S$_N$2 mechanism (or an intermediate pathway with both S$_N$1 and S$_N$2 character).
- *Tertiary* alkyl halides are likely to react by an S$_N$1 mechanism.

The nucleophile

- Increasing the nucleophilic strength of the nucleophile *will* increase the rate of an S$_N$2 reaction.
- Increasing the nucleophilic strength of the nucleophile *will not* increase the rate of an S$_N$1 reaction.

The mechanism can shift from S$_N$1 to S$_N$2 on changing to a more powerful nucleophile.

The solvent

- Increasing the polarity (i.e. the dielectric constant, ε) of the solvent will result in a *slight decrease* in the rate of an S$_N$2 reaction. This is because the charge is more spread out in the transition state than in the reactants.
- Increasing the polarity of the solvent will result in a *significant increase* in the rate of an S$_N$1 reaction. This is because polar solvents, with high dielectric constants (e.g. water and methanol), can

stabilise the carbocation intermediate (formed in an S_N1 reaction) by solvation. In an S_N2 reaction, solvation is likely to stabilise the attacking nucleophile (which has a more concentrated negative charge) rather than the transition state (which has a less concentrated charge). The mechanism can therefore shift from S_N2 to S_N1 on changing to a more polar solvent.

- In addition, the nucleophilic strength of a nucleophile depends on whether a polar protic solvent or a polar non-protic (or aprotic) solvent is used. This is because polar protic solvents (e.g. methanol) can stabilise and therefore lower the reactivity of the attacking nucleophile, by hydrogen bonding. The nucleophilic strength is increased in polar non-protic solvents (e.g. dimethylsulfoxide, Me_2SO), because these are not as effective at solvating the nucleophile (i.e. they are not able to hydrogen bond to a negatively charged nucleophile). The mechanism can therefore shift from S_N1 to S_N2 on changing from a polar protic solvent to a polar non-protic solvent.

5.3.1.4 The leaving group

The better the leaving group, the faster the rate of *both* S_N1 and S_N2 reactions (as C–X bond cleavage is involved in both of the rate-determining steps). The best leaving groups are those that form stable neutral molecules or stable anions, which are weakly basic. The less basic the anion, the more stable (or less reactive) the anion.

Leaving group ability	H_2O > $HO^{\ominus}$
	$I^{\ominus}$ > $Br^{\ominus}$ > $Cl^{\ominus}$ > $F^{\ominus}$
	weak base *strong base*

Amongst the halogens, the iodide anion (I^-) is the best leaving group, as it forms a weak bond to carbon. The C–I bond is therefore relatively easily broken to give I^-. I^- is a weak base, and weak bases are best able to accommodate the negative charge (i.e. weak bases are the best leaving groups).

5.3.1.5 Nucleophilic catalysis

The iodide anion (I^-) is a good nucleophile and a good leaving group. It can be used as a catalyst to speed up the rate of a slow S_N2 reaction.

5.3.1.6 Tight ion pairs

In practice, few S_N1 reactions lead to complete racemisation because of the formation of *tight* (or *intimate*) *ion pairs*. The carbocation and negatively charged leaving group interact so that the negatively charged nucleophile enters predominantly from the side opposite to the departing leaving group. This results in a greater proportion of inversion.

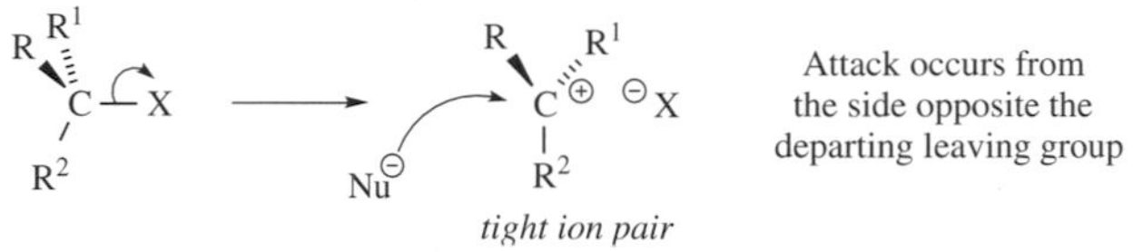

Attack occurs from the side opposite the departing leaving group

tight ion pair

5.3.1.7 The S_Ni reaction (substitution, nucleophilic, internal reaction)

The formation of a tight (or intimate) ion pair can also lead to the retention of configuration. This is observed when alcohols with asymmetric centres react with thionyl chloride ($SOCl_2$) in the absence of a nitrogen base. The intermediate alkyl chlorosulfite (ROSOCl) breaks down to form a tight ion pair (R^+ ^-OSOCl) within a solvent cage. The ^-OSOCl anion then collapses to give SO_2 and Cl^-, which attack the carbocation from the same side that ^-OSOCl departs.

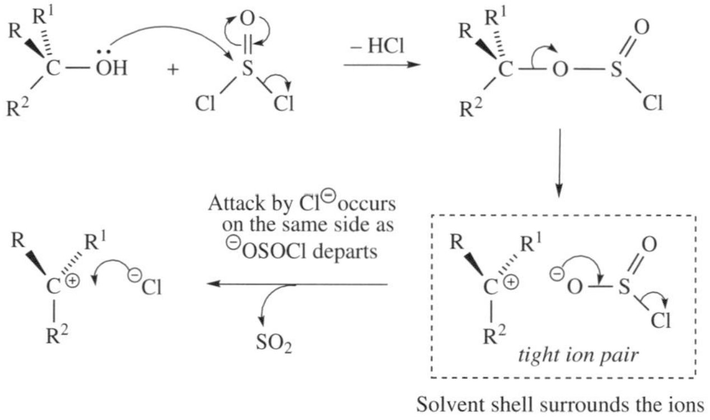

Attack by $Cl^\ominus$ occurs on the same side as $^\ominus OSOCl$ departs

tight ion pair

Solvent shell surrounds the ions

- In the *absence* of a nitrogen base, HCl is formed, and this cannot act as a nucleophile and attack the alkyl chlorosulfite in an S_N2 reaction. Therefore, an S_Ni mechanism operates.
- An S_N2 reaction is observed in the *presence* of a nitrogen base (B) because the HCl is converted to BH^+ Cl^-. The Cl^- can then act as a nucleophile and attack the alkyl chlorosulfite in an S_N2 reaction (see Section 5.2.2).

5.3.1.8 Neighbouring group participation (or anchimeric assistance)

Two consecutive S_N2 reactions will lead to a retention of configuration (i.e. two inversions = retention). This can occur when a neighbouring group acts as a nucleophile in the first of two S_N2 reactions.

- The *first* S_N2 reaction is an *intramolecular* reaction (i.e. a reaction within the same molecule).
- The *second* S_N2 reaction is an *intermolecular* reaction (i.e. a reaction between two different molecules).

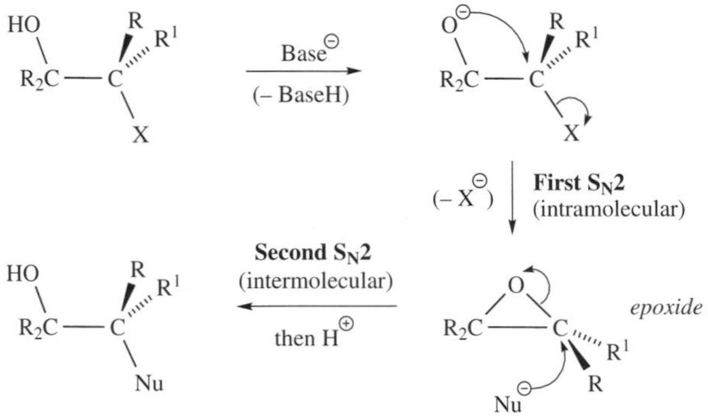

The presence of neighbouring groups, including R_2N and RS, can increase the rate of substitution reactions.

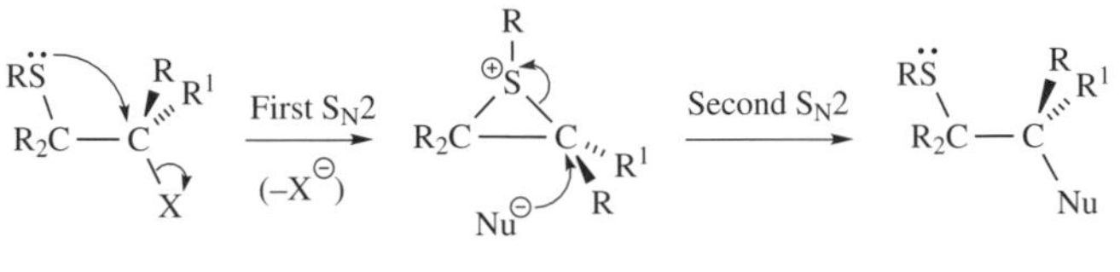

5.3.1.9 The S_N2' and S_N1' reactions

Allylic halides can undergo substitution with an allylic rearrangement (i.e. change in the position of the double bond) in one of the two ways.

S_N2' mechanism: by nucleophilic attack on the double bond

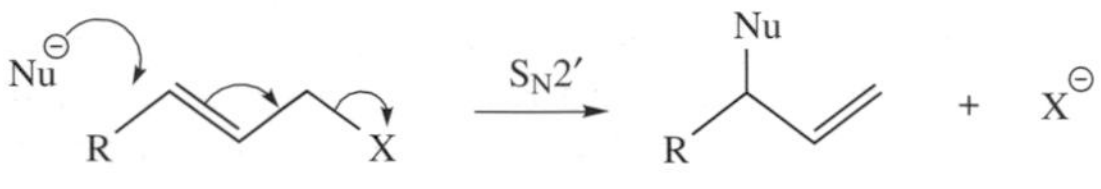

S_N1' mechanism: by nucleophilic attack on a carbocation resonance structure

S_N1' mechanism: by nucleophilic attack on a carbocation resonance structure

primary carbocation

secondary carbocation

5.3.2   Elimination

The two main mechanisms for elimination reactions of alkyl halides are E1 and E2. In both cases, the mechanism involves the loss of HX from RX to give an alkene.

5.3.2.1   The E2 (elimination, bimolecular) reaction

This is a concerted (one-step) mechanism. The C=C bond begins to form at the same time as the C–H and C–X bonds begin to break. The reaction is second order (or bimolecular), as the rate depends on the concentration of the base and the alkyl halide.

$$\text{Reaction rate} = k[\text{Base}][\text{RX}]$$

The elimination requires the alkyl halide to adopt an *antiperiplanar* shape (or conformation), in which the H and X groups are on the opposite sides of the molecule. (*Synperiplanar* conformation is when the H and X groups are on the same side of the molecule.) The antiperiplanar arrangement is lower in energy than the synperiplanar arrangement, as this has a staggered, rather than an eclipsed, conformation.

The H–C–C–X atoms are in the same plane (i.e. periplanar)

Transition state

Antiperiplanar (staggered)

$$\text{Base} - \text{H} \quad + \quad \overset{R^1}{\underset{R}{>}}\text{C}=\text{C}\overset{R^2}{\underset{R^3}{<}} \quad + \quad \text{X}^{\ominus}$$

The elimination is *stereospecific*, as different stereoisomers of the alkyl halide give different stereoisomers of the alkene. This is because the new π-bond is formed by the overlap of the C–H σ-bond with the C–X σ*-bond. These orbitals must be in the same plane for the best overlap, and they become p-orbitals in the π-bond of the alkene.

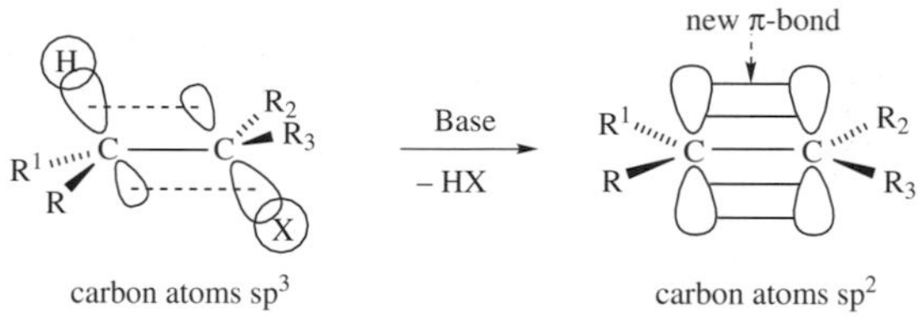

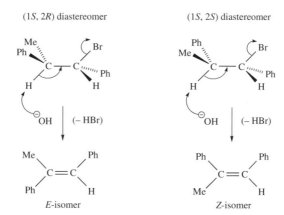

carbon atoms sp^3 carbon atoms sp^2

The antiperiplanar conformation is responsible for:

(1) the stereospecific formation of substituted alkenes (i.e. *E* or *Z* configuration);
(2) the different rates of elimination of HX from cyclohexyl halide conformers.

Elimination reactions of alkyl halide diastereoisomers

A different alkene stereoisomer is obtained from each diastereoisomer of the alkyl halide.

(1*S*, 2*R*) diastereomer   (1*S*, 2*S*) diastereomer

E-isomer   *Z*-isomer

Elimination reactions of cyclohexyl halide conformers

For elimination in cyclohexanes, both the C–H and C–X bonds must be axial.

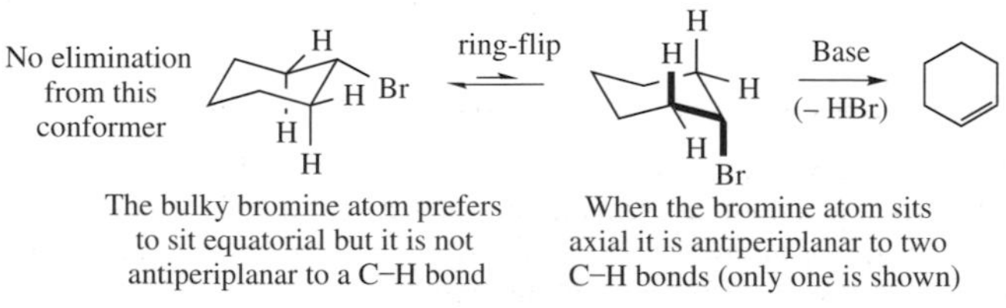

No elimination from this conformer   ring-flip   Base (– HBr)

The bulky bromine atom prefers to sit equatorial but it is not antiperiplanar to a C–H bond

When the bromine atom sits axial it is antiperiplanar to two C–H bonds (only one is shown)

Regioselectivity

The regioselectivity (i.e. the position of the double bond in the alkene) of the E2 reaction depends on the nature of the leaving group (X).

| If the C–X bond is not easily broken then a carbanion-like transition state will lead to the least stable (or substituted) alkene | If the C–X bond is easily broken then an alkene-like transition state will lead to the most stable (or substituted) alkene |

$$H_B \underset{R^1}{\overset{}{\underset{R}{}}}C - \underset{X}{\overset{H}{C}} - C\overset{H_A}{\underset{H}{}}H$$

Base $\diagup$ $-H_AX$ $-H_BX$ $\diagdown$ Base

$X = \overset{\oplus}{N}Me_3$ $X = Br$

carbanion-like transition state   alkene-like transition state

least substituted alkene
Hofmann elimination

most substituted alkene
Saytzev elimination

- *When X = Br.* The elimination reaction forms predominantly the *most* highly substituted alkene in a *Saytzev* (or *Zaitsev*) *elimination*. The H–C bond is broken at the same time as the C–Br bond, and the transition state resembles the alkene product. As a result, the more highly substituted alkene is formed faster, because this is the more stable alkene product (see Section 6.1).

- *When X = +NMe₃.* The elimination reaction forms predominantly the *least* highly substituted alkene in a *Hofmann elimination*. The H–C bond is broken before the C–+NMe₃ bond, and the transition state resembles a carbanion product. As a result, the base removes H^A to form a more stable carbanion (i.e. the δ– is situated on a carbon bearing the least number of electron-donating alkyl groups). The Hofmann elimination is also normally expected when using bulky bases. This is because H^A is more accessible (i.e. less hindered and more easily attacked) than H^B.

5.3.2.2 The E1 (elimination, unimolecular) reaction

This is a stepwise mechanism involving an initial cleavage of the carbon–halogen bond to form an intermediate carbocation. The reaction is first order, or unimolecular, as the rate depends on the concentration of only the alkyl halide (and not the base).

$$\text{Reaction rate} = k[RX]$$

The reaction does not require a particular geometry; the intermediate carbocation can lose any proton from an adjacent position.

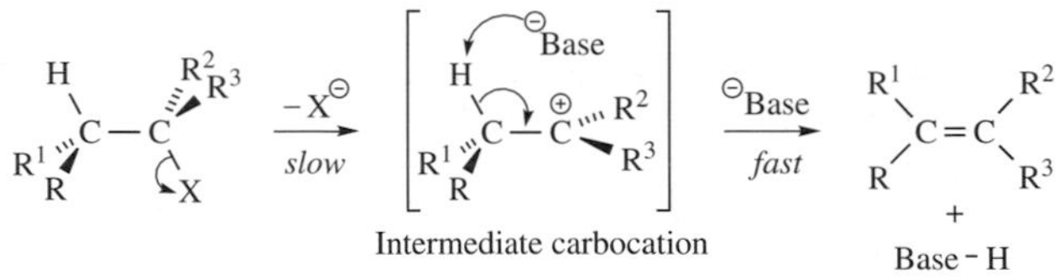

Intermediate carbocation

Base – H

Tertiary halides react more rapidly than primary halides, because the intermediate carbocation is more stable and therefore more readily formed.

Order of E1 reactivity = tertiary > secondary > primary

E1 reactions can be regio- and stereoselective.

Regioselectivity

E1 eliminations give predominantly the more stable (or more substituted) alkene. The transition state (for the loss of a proton from the intermediate carbocation) leading to the more substituted alkene will be lower in energy.

Stereoselectivity

E1 eliminations give predominantly the E-alkene rather than the Z-alkene. The E-alkene is more stable than the Z-alkene for steric reasons (i.e. the bulky substituents are further apart in E-alkenes). The transition state leading to E-alkenes will therefore be lower in energy.

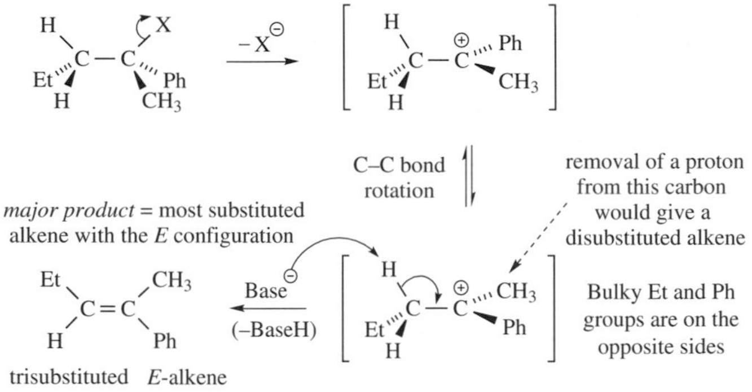

C–C bond rotation

removal of a proton from this carbon would give a disubstituted alkene

major product = most substituted alkene with the E configuration

trisubstituted E-alkene

Bulky Et and Ph groups are on the opposite sides

5.3.2.3   E2 reactions versus E1 reactions

The mechanism of an elimination reaction is influenced by the nature of the alkyl halide, the base and the solvent.

The alkyl halide

- Primary, secondary and tertiary alkyl halides can all undergo an E2 mechanism (hence, this is more common than E1).
- Tertiary alkyl halides can react by an E1 mechanism.
- The better the leaving group, the faster the rate of both E1 and E2 reactions (as C–X bond cleavage is involved in both of the rate-determining steps).

The base

- Increasing the strength of the base *will* increase the rate of an E2 elimination. Strong bases such as HO⁻ and RO⁻ will favour an E2 elimination.
- Increasing the strength of the base *will not* increase the rate of an E1 reaction.
- The mechanism can shift from E1 to E2 on changing to a more powerful base.

The solvent

- Increasing the polarity (i.e. the dielectric constant, ε) of the solvent will result in a *slight decrease* in the rate of an E2 reaction.
- Increasing the polarity of the solvent will result in a *significant increase* in the rate of an E1 reaction.
  This is because polar solvents, with high dielectric constants (e.g. water and methanol), can stabilise the carbocation intermediate (formed in an E1 reaction) by solvation. For an E2 reaction, solvation is likely to stabilise the attacking base (which has a more concentrated negative charge) rather than the transition state (which has a less concentrated charge). The mechanism can therefore shift from E2 to E1 on changing to a more polar solvent.
- The strength of a base depends on whether a polar protic solvent or a polar non-protic (or aprotic) solvent is used. This is because polar protic solvents (e.g. methanol) can stabilise and therefore lower the reactivity of the attacking base, by hydrogen bonding. The basic strength is increased in polar non-protic solvents (e.g. dimethylsulfoxide, Me$_2$SO), because these are not as effective at solvating the base (i.e. they are not able to hydrogen bond to a negatively charged base). The mechanism can therefore shift from E1 to E2 on changing from a polar protic solvent to a polar non-protic solvent.

5.3.2.4 The E1cB (elimination, unimolecular, conjugate base) reaction

This is a stepwise mechanism involving an initial deprotonation of the alkyl halide to form an intermediate carbanion. The carbanion (or the conjugate base of the starting material) then loses the leaving group in a slow (rate-determining) step.

Intermediate carbanion

The reaction is favoured by the presence of electron-withdrawing (−I, −M) groups on the β-carbon atom (i.e. R and R^1 = electron withdrawing). Suitable groups include RC=O and RSO_2, which can stabilise the intermediate carbanion. These groups therefore increase the acidity of the β-hydrogen atoms, leading to an initial deprotonation.

5.3.3   Substitution versus elimination

Alkyl halides undergo competitive substitution and elimination reactions. The ratio of products derived from substitution and elimination depends on the nature of the alkyl halide, the base/nucleophile, the solvent and the temperature. S_N2 reactions are normally in competition with E2 reactions, while S_N1 reactions are normally in competition with E1 reactions.

S_N2 versus E2

Primary and secondary alkyl halides can undergo S_N2 reactions, while primary, secondary and tertiary alkyl halides can undergo E2 reactions.

For primary and secondary halides

- S_N2 reactions are favoured over E2 reactions when using strong nucleophiles in polar aprotic solvents.
- E2 reactions are favoured over S_N2 reactions by using strong bulky bases (which are poor nucleophiles).

In general, large nucleophiles are good bases and promote elimination. This is because the bulky anion cannot attack the hindered carbon atom in an S_N2 reaction. It is much easier for the anion to attack a β-hydrogen atom because this is more accessible.

With a large anion, this is more likely to act as a base and react with the hydrogen on the β-carbon because this is more exposed than the α-carbon atom

Order of basicity $\quad H_3C-\underset{\underset{CH_3}{|}}{\overset{\overset{CH_3}{|}}{C}}-O^{\ominus} \quad > \quad H_3C-O^{\ominus}$

E2 reactions are favoured over S_N2 reactions by using high temperatures. In general, increasing the reaction temperature leads to more elimination. This is because the elimination reaction has a higher activation energy (than the substitution reaction), as more bonds need to be broken in order to form the alkene product.

It should also be noted that in an elimination reaction, two molecules react to give three new molecules. In contrast, for a substitution reaction, two molecules react to form two new molecules. The change in entropy ($\Delta S°$) is therefore greater for an elimination reaction. From the Gibbs free energy equation (see Section 4.9.1), $\Delta S°$ is multiplied by the temperature (T), and the larger the $T\Delta S°$ term, the more favourable $\Delta G°$.

S_N1 versus E1

Secondary and tertiary alkyl halides can undergo S_N1 or E1 reactions. Allylic and benzylic halides can also undergo S_N1 or E1 reactions. (Tertiary halides usually undergo E2 elimination in the presence of strong bases.) Both S_N1 and E1 reactions can occur when using weakly basic or non-basic nucleophiles in protic solvents.

E1 reactions are favoured over S_N1 reactions by using high temperatures. E1 reactions are also favoured as the size of the alkyl groups (on the α-carbon atom) increases. The bigger these groups are, the harder it is for the nucleophile to attack the α-carbon atom.

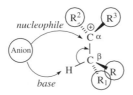

The bigger the alkyl groups the more likely the anion will attack the hydrogen atom on the β-carbon

$S_N2/E2$ versus $S_N1/E1$

* $S_N1/E1$ reactions are favoured over $S_N2/E2$ reactions by using polar protic solvents (which can solvate the carbocation intermediates).
* $S_N2/E2$ reactions are favoured over $S_N1/E1$ reactions by using strong nucleophiles or bases.
* $S_N2/E2$ reactions are favoured over $S_N1/E1$ reactions by using high concentrations of the nucleophile/base (as the rate of these bimolecular reactions depends on the concentration of the nucleophile or base).

Halide	Mechanism
Primary: RCH_2X	$\mathbf{S_N2}$, *E2*
Secondary: R_2CHX	S_N1, $\mathbf{S_N2}$, E1, *E2*
Tertiary: R_3CX	S_N1, E1, *E2*

In *italic* = favoured when using a strong base

In **bold** = favoured when using a good nucleophile

Problems

(1) For the reaction of compound **A** with bromide ion in propanone (acetone):

$$\underset{\textstyle \text{A}}{\overset{\textstyle \text{Et}}{\underset{\textstyle \text{Me}}{\overset{|}{\underset{H^{\prime\prime\prime\prime}}{C}}}}\diagdown I}$$

(a) Draw the product of an S_N2 reaction.
(b) Predict what other substitution product might be formed if the reaction is carried out in ethanol instead of propanone.
(c) Explain why, if **A** is reacted with hydroxide ion rather than bromide ion, little substitution product results.
(d) Explain why, in an S_N1 reaction, compound **A** reacts less rapidly than does (1-iodoethyl)benzene, $PhCH(I)CH_3$.

(2) Hydrolysis of 2-bromopentane (**B**) can occur by either an S_N1 pathway or an S_N2 pathway, depending on the solvent and conditions employed.

(a) Draw the structure of the (*R*)-isomer of **B**.
(b) Write down the S_N1 and S_N2 mechanism of the above reaction, starting from the (*R*)-isomer of **B**.
(c) Suggest, giving your reasons, conditions under which the reaction might be expected to proceed by an S_N2 route.
(d) Explain why the addition of catalytic amounts of sodium iodide accelerates the formation of **C** from **B** under S_N2 but not S_N1 conditions.
(e) What product(s) would you expect from the reaction of **B** with sodium *tert*-butoxide, $Me_3C-O^-\ Na^+$?
(f) Explain why **C** is stable in ethanol, but if a small amount of sulfuric acid is added, it is slowly converted to 2-ethoxy-pentane.

(3) When the solvent acts as the nucleophile in a substitution reaction, this is known as *solvolysis*. What products would you expect from the solvolysis of $PhCH_2Br$ in (a) methanol and (b) ethanoic acid?

(4) When (*R*)-butan-2-ol is reacted with excess $SOCl_2$, the main product is (*R*)-2-chlorobutane. However, if the same is carried out in the presence of pyridine, the major product is (*S*)-2-chlorobutane. Why do the two reactions give different products?

(5) What product would be formed on E2 elimination of (1*S*,2*S*)-1,2-dichloro-1,2-diphenylethane?

(6) Reaction of bromide **D** in hot ethanol gave a number of products including **E** and **F**. Suggest reaction mechanisms to explain the formation of **E** and **F**.

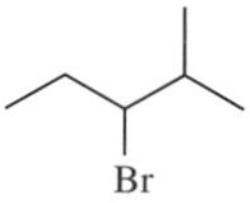

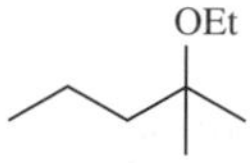

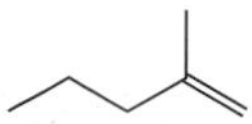

6. ALKENES AND ALKYNES

Key point. Alkenes and alkynes are unsaturated hydrocarbons, which possess a C=C double bond and a C≡C triple bond, respectively. As (weak) π-bonds are more reactive than (strong) σ-bonds, alkenes and alkynes are more reactive than alkanes. The electron-rich double or triple bond can act as a nucleophile, and most reactions of alkenes/ alkynes involve *electrophilic addition reactions*. In these reactions, the π-bond attacks an electrophile to generate a carbocation, which then reacts with a nucleophile. Overall, these reactions lead to the addition of two new substituents at the expense of the π-bond.

6.1 Structure

Alkenes have a C=C double bond. The two carbon atoms in a double bond are sp^2 hydridised, and the C=C bond contains one (strong) σ-bond and one (weaker) π-bond.

Alkynes have a C≡C triple bond. The two carbon atoms in a triple bond are sp hybridised, and the C≡C bond contains one σ-bond and two π-bonds.

Alkene (planar shape)   Alkyne (linear shape)

The greater the number of π-bonds, the shorter and stronger the carbon–carbon bond becomes.

	HC≡CH		H$_2$C=CH$_2$		H$_3$C−CH$_3$
CC bond strength (kJ mol^{-1})	837	>	636	>	368
CC bond length (Å)	1.20	<	1.33	<	1.54

As there is no free rotation around a C=C double bond, alkenes can have *E*- and *Z*-stereoisomers (see Section 3.3.1). *Z*-Isomers are less stable than *E*-isomers because of the steric interactions between the two bulky groups on the same side of the molecule.

steric strain

Z-isomer   E-isomer
less stable   **more stable**

An alkene is stabilised by alkyl substituents on the double bond. When the stability of alkene isomers is compared, it is found that the greater the number of alkyl substituents on the C=C double bond, the more stable the alkene is. As an approximation, it is the number of alkyl substituents rather than their identities that govern the stability of an alkene. This can be explained by hyperconjugation which involves the donation of electrons from a filled C–H σ-orbital to an empty C=C π^*-orbital.

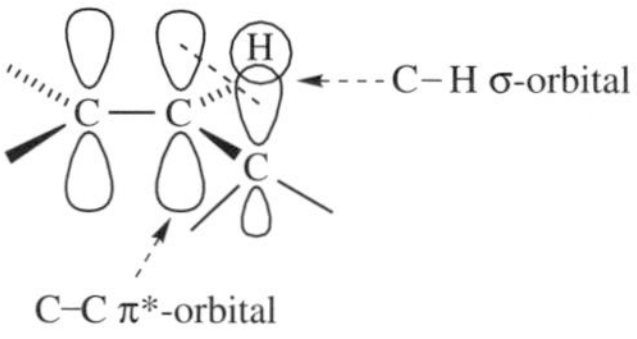

C–C π^*-orbital

←---C–H σ-orbital

Overlap between these orbitals leads to the stabilisation of the double bond. The more alkyl substituents on the double bond, the greater the number of C–H bonds which can interact, and hence the more stable the alkene is

most stable   **least stable**

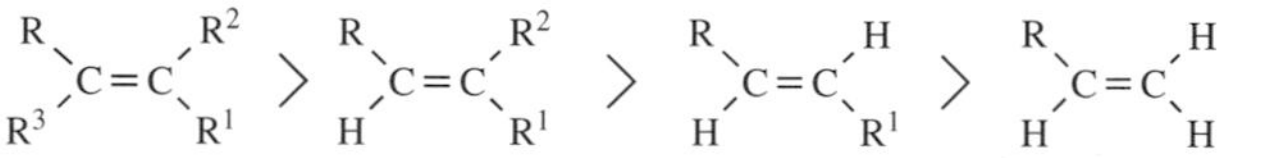

tetrasubstituted   trisubstituted   disubstituted   monosubstituted

It should be emphasised that both alkenes and alkynes have electron-rich π-bonds and hence they react as nucleophiles.

6.2 Alkenes

6.2.1 Preparation

An important method for preparing alkenes involves the elimination (E1 or E2) of HX from alkyl halides (see Section 5.3.2). Alcohols can also be converted to alkenes by activating the OH group (e.g. by protonation or conversion to a tosylate) to make this into a better leaving group (see Section 5.2.2).

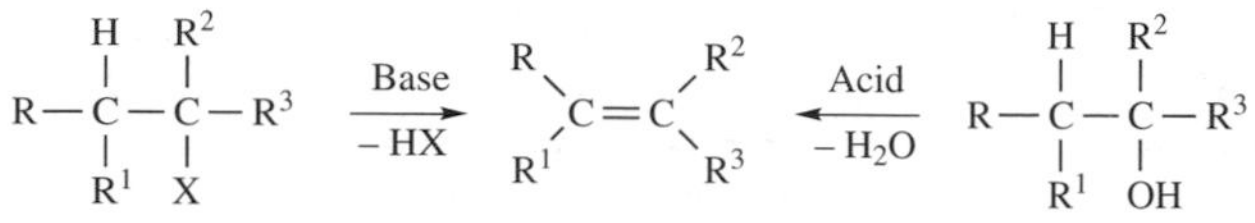

Tertiary amine oxides and xanthates can also undergo elimination to form alkenes on heating. These *Ei eliminations* (*elimination, intramolecular*) proceed *via* cyclic transition states, which require a synperiplanar arrangement of the leaving groups (i.e. the leaving groups lie in the same plane on the same side of the molecule). This can be compared to an E2 elimination, which requires an antiperiplanar arrangement of leaving groups (see Section 5.3.2.1).

The Cope elimination

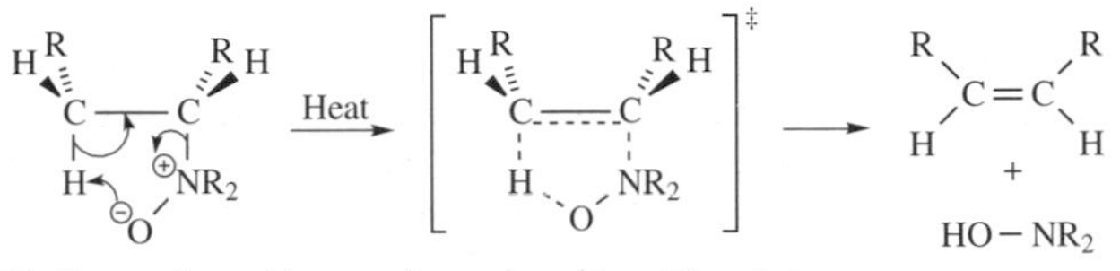

Tertiary amine oxide   *5-membered transition state*

The Chugaev elimination

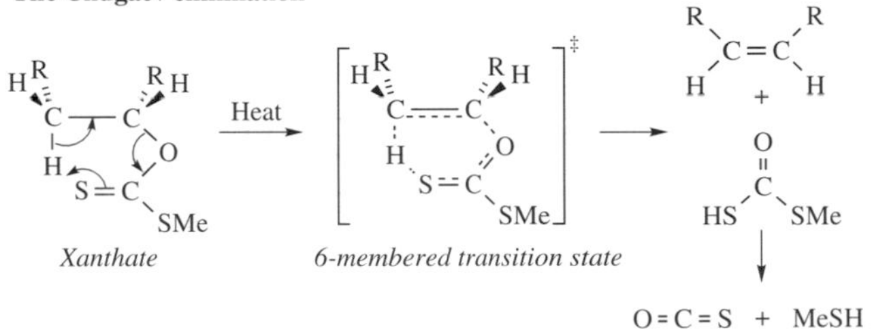

Xanthate   *6-membered transition state*

$$O = C = S \ + \ MeSH$$

Alkenes can also be prepared using the Wittig reaction (see Section 8.3.4.3), and Z-alkenes can be prepared by the hydrogenation of alkynes in the presence of a Lindlar catalyst (see Section 6.3.2.4).

6.2.2 Reactions

Alkenes are nucleophiles and react with electrophiles in *electrophilic addition reactions*. These reactions lead to the introduction of two new substituents at the expense of the π-bond. (Some of these reactions were introduced in Section 5.2.3.)

For unsymmetrical alkenes, with different substituents at either end of the double bond, the electrophile adds regioselectively, so as to form the more substituted (and therefore the more stable) carbocation.

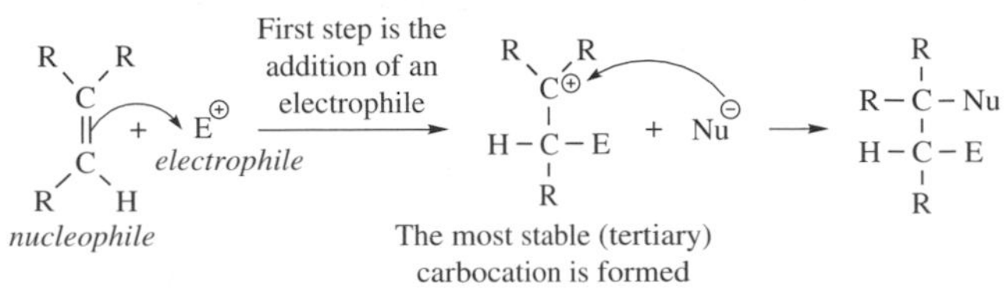

The more alkyl groups on the double bond, the faster the rate of electrophilic addition. This is because the electron-donating (+I) alkyl groups make the double bond more nucleophilic (e.g. $Me_2C=CMe_2$ is much more reactive to electrophiles than $H_2C=CH_2$). Conversely, the introduction of electron-withdrawing (−I, −M) substituents reduces the rate (e.g. $H_2C=CH-CH_3$ is much more reactive to electrophiles than $H_2C=CH-CF_3$).

6.2.2.1 Addition of hydrogen halides

The addition of HX (X=Cl, Br, I) to an alkene, to form alkyl halides, occurs in two steps. The first step involves the addition of a proton (i.e. the electrophile) to the double bond to make the most stable intermediate carbocation. The second step involves nucleophilic attack by the halide anion. This gives a racemic alkyl halide product because the carbocation is planar and hence can be attacked equally from either face. (These addition reactions are the reverse of alkyl halide elimination reactions.)

The proton adds to the less substituted carbon atom

tertiary carbocation

The tertiary rather than secondary carbocation is formed because this is more stable

secondary carbocation

The regioselective addition of HX to alkenes produces the more substituted alkyl halide, which is known as the Markovnikov (Markovnikoff) product. Markovnikov's rule states that: 'on addition of HX to an alkene, H attaches to the carbon with fewest alkyl groups and X attaches to the carbon with most alkyl groups'. This can be explained by the formation of the most stable intermediate carbocation.

Occasionally, the intermediate carbocations can also undergo structural rearrangements to form more stable carbocations. A hydrogen atom (with its pair of electrons) can migrate onto an adjacent carbon atom in a '*hydride shift*'. This will only occur if the resultant cation is more stable than the initial cation.

regioselective addition of a proton to give the secondary, rather than primary, carbocation

hydride shift

This produces a more stable tertiary carbocation

Alkyl groups can also migrate to a carbocation centre in *Wagner–Meerwein rearrangements* (or *shifts*).

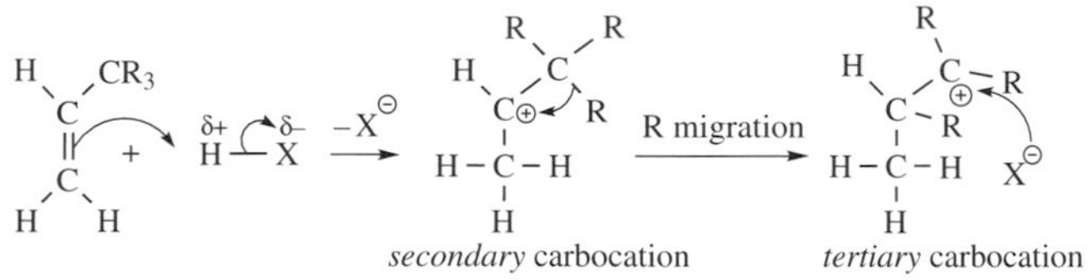

Addition of HBr to alkenes in the presence of a peroxide

Although HBr usually adds to alkenes to give the Markovnikov product, in the presence of a peroxide (ROOR), HBr adds to alkenes to give the alternative regioisomer. This is known as the anti-Markovnikov product. The change in regioselectivity occurs because the mechanism of the reaction changes from an ionic (polar) mechanism in the absence of peroxides to a radical mechanism in the presence of peroxides. Initiation by peroxide leads to the formation of bromine radicals (or atoms) which add to the less hindered end of the alkene. This gives the more substituted radical, which in turn produces the less substituted alkyl halide. (It should be remembered that radical stability follows the same order as cation stability – see Section 4.3.)

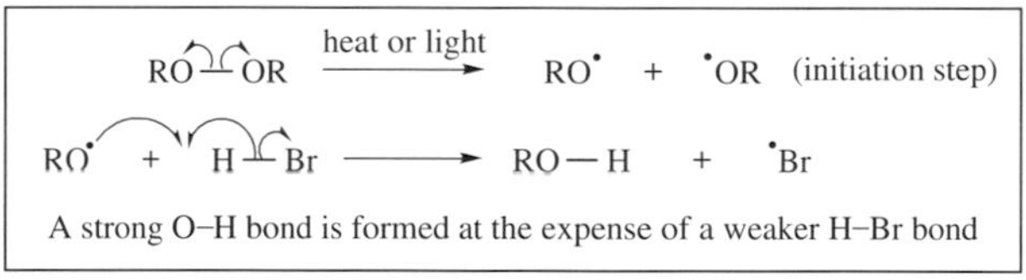

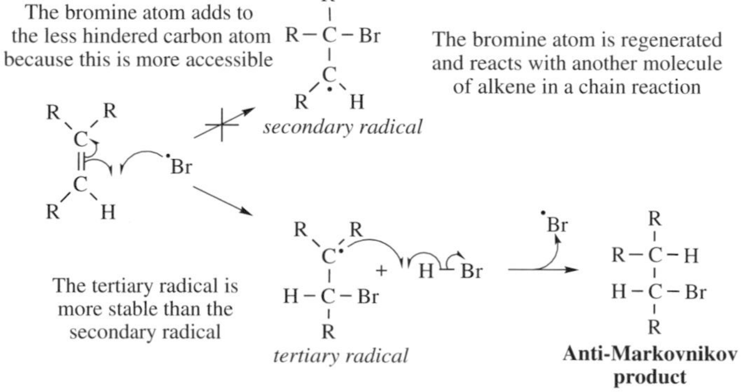

6.2.2.2   Addition of bromine

The electrophilic addition of bromine produces 1,2-dibromide (or vicinal dibromide). The addition is stereospecific because of the formation of an intermediate bromonium ion. This ensures that the bromine atoms add to the opposite sides of the alkene in an *anti-* addition.

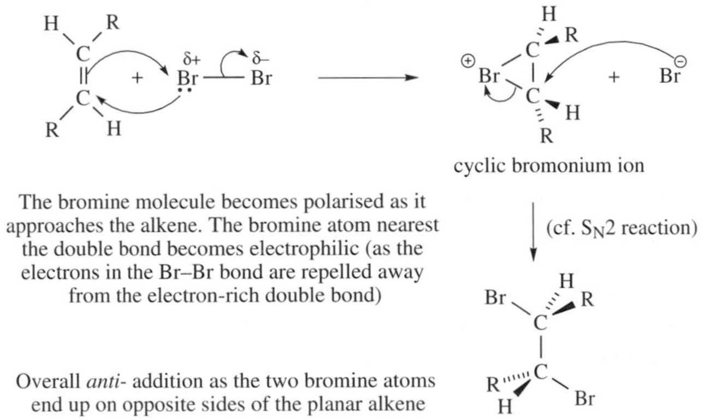

cyclic bromonium ion

The bromine molecule becomes polarised as it approaches the alkene. The bromine atom nearest the double bond becomes electrophilic (as the electrons in the Br–Br bond are repelled away from the electron-rich double bond)

(cf. S_N2 reaction)

Overall *anti-* addition as the two bromine atoms end up on opposite sides of the planar alkene

The *anti-* addition explains the formation of different diastereoisomers when using *E*- or *Z*-alkenes.

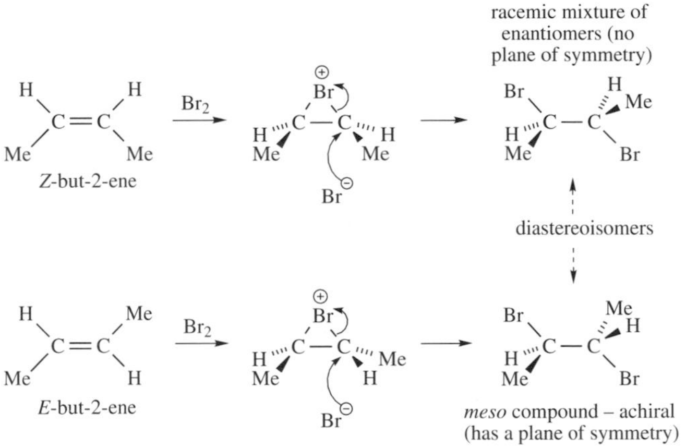

racemic mixture of enantiomers (no plane of symmetry)

Z-but-2-ene

diastereoisomers

E-but-2-ene

meso compound – achiral (has a plane of symmetry)

As Br$^-$ can attack either carbon atom of the bromonium ion, reaction with *Z*-but-2-ene produces a 1:1 mixture of enantiomers (only the 2*R*,3*R* isomer is shown above). For *E*-but-2-ene, the attack of Br$^-$ at either carbon atom of the bromonium ion produces the same compound. This compound has a plane of symmetry and hence is an achiral *meso* compound.

   N-Bromosuccinimide or NBS (a stable, easily handled solid), rather than hazardous liquid bromine, is often used in these bromination reactions, as this produces bromine (at a controlled rate) on reaction with HBr (which is usually present in trace amounts in NBS).

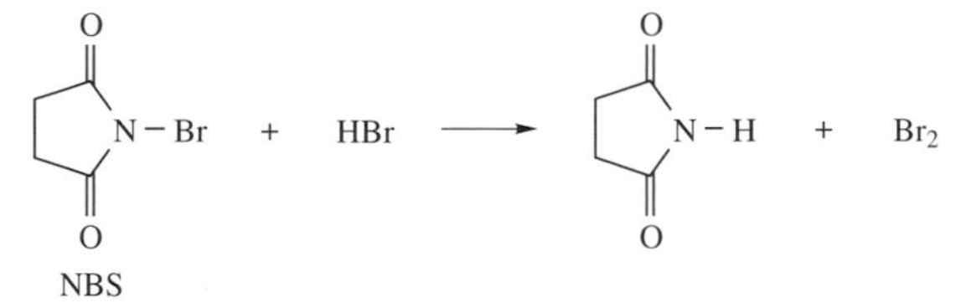

$$N-Br \;+\; HBr \longrightarrow N-H \;+\; Br_2$$

NBS

6.2.2.3   Addition of bromine in the presence of water

The addition of bromine to an alkene in the presence of water can lead to the formation of a 1,2-bromo-alcohol (or bromohydrin) in addition to a 1,2-dibromide. This is because water can act as a nucleophile and compete with the bromide ion for ring-opening of the bromonium ion. Even though Br⁻ is a better nucleophile than H_2O, if H_2O is present in excess, then the 1,2-bromo-alcohol will be formed.

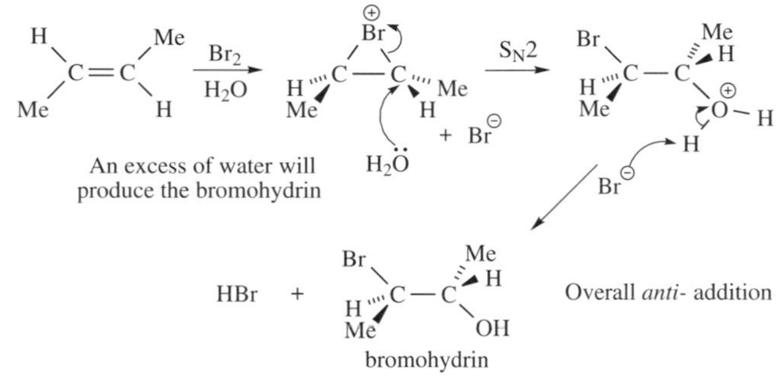

An excess of water will produce the bromohydrin

S_N2

$+$ Br⁻

H_2O

Overall *anti*- addition

HBr +

bromohydrin

The opening of the bromonium ion is often regioselective. The nucleophile usually attacks the more substituted carbon atom of the ring, because this carbon atom is more positively polarised. The reaction proceeds via a 'loose S_N2 transition state'.

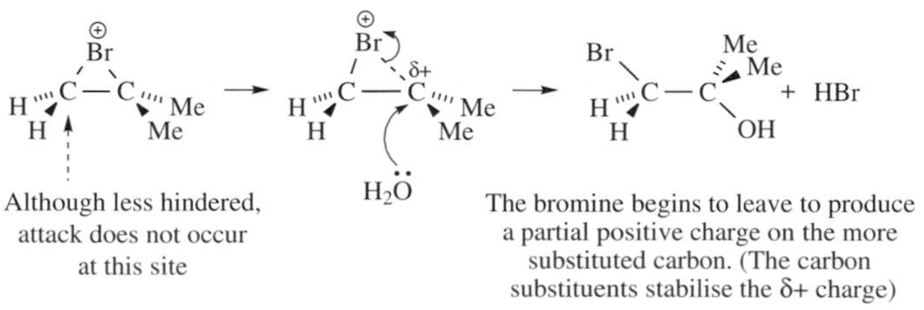

Although less hindered, attack does not occur at this site

H_2O

+ HBr

The bromine begins to leave to produce a partial positive charge on the more substituted carbon. (The carbon substituents stabilise the δ+ charge)

6.2.2.4 Addition of water (hydration): Markovnikov addition

The addition of water to alkenes, to produce alcohols, requires the presence of either (i) a strong acid or (ii) mercury(II) acetate (in an oxymercuration reaction). In both cases, the reactions involve the Markovnikov addition of water (i.e. the OH becomes attached to the more substituted carbon).

Acid catalysis (requires high temperatures)

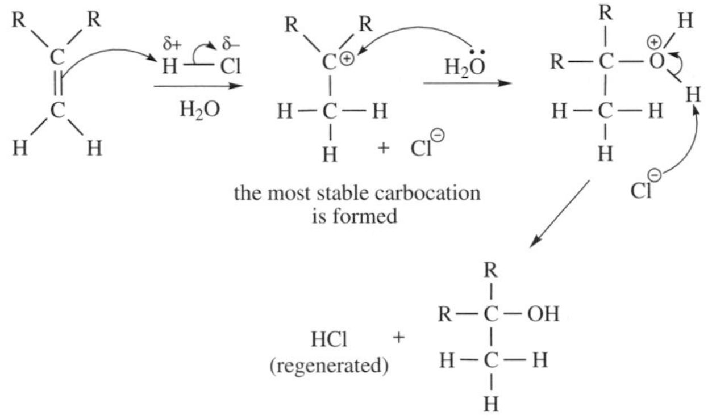

the most stable carbocation
is formed

HCl   +
(regenerated)

Oxymercuration

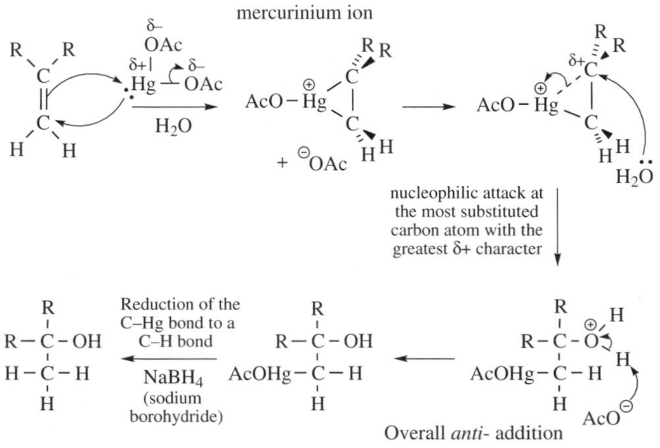

mercurinium ion

nucleophilic attack at
the most substituted
carbon atom with the
greatest δ+ character

Reduction of the
C–Hg bond to a
C–H bond

NaBH₄
(sodium
borohydride)

Overall *anti-* addition

6.2.2.5 Addition of water (hydration): anti-Markovnikov addition

The anti-Markovnikov addition of water is achieved using borane (B_2H_6, which reacts as BH_3) in a hydroboration reaction. The reaction involves the *syn-* addition of a boron–hydrogen bond to the alkene *via* a 4-membered transition state (i.e. the boron atom and the hydrogen atom add to the same face of the alkene). Hydroboration is a highly regioselective reaction, and steric factors are important. The boron atom adds to the least hindered end of the alkene to give an organoborane (regioselectively), which is then oxidised to alcohol. Bulkier boranes such as 9-BBN (9-borabicyclo[3.3.1]nonane) can enhance the regioselectivity of hydroboration.

Hydroboration (regioselective and stereoselective)

| The boron atom adds to the less hindered carbon atom | 4-membered transition state | *syn-* addition: the H and BH_2 add to the same face of the alkene |

Electronic factors also play a part. Boron is more electropositive than hydrogen, and hence the double bond will attack the boron atom to give a partial positive charge on one of the (alkene) carbon atoms in the 4-membered transition state. The partial positive charge will reside on the more substituted carbon atom, as this is more stable.

In the presence of two further alkene molecules, a trialkylborane is formed. Two further alkyl groups replace the two hydrogen atoms on boron (using the same mechanism as shown above).

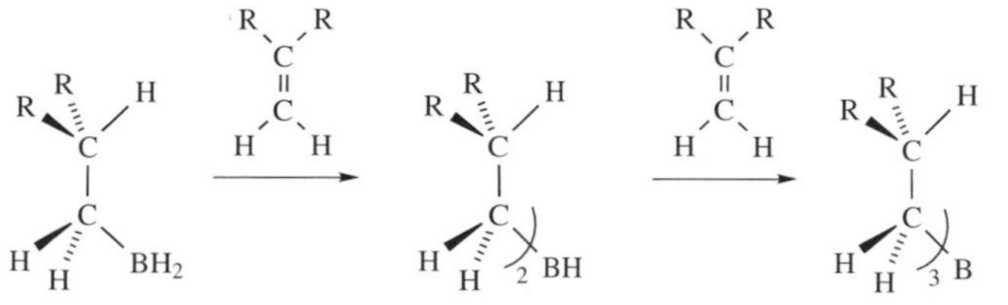

trialkylborane

The trialkylborane is then oxidised using hydrogen peroxide (H_2O_2) in basic solution. This converts the three C–B bonds to C–OH bonds.

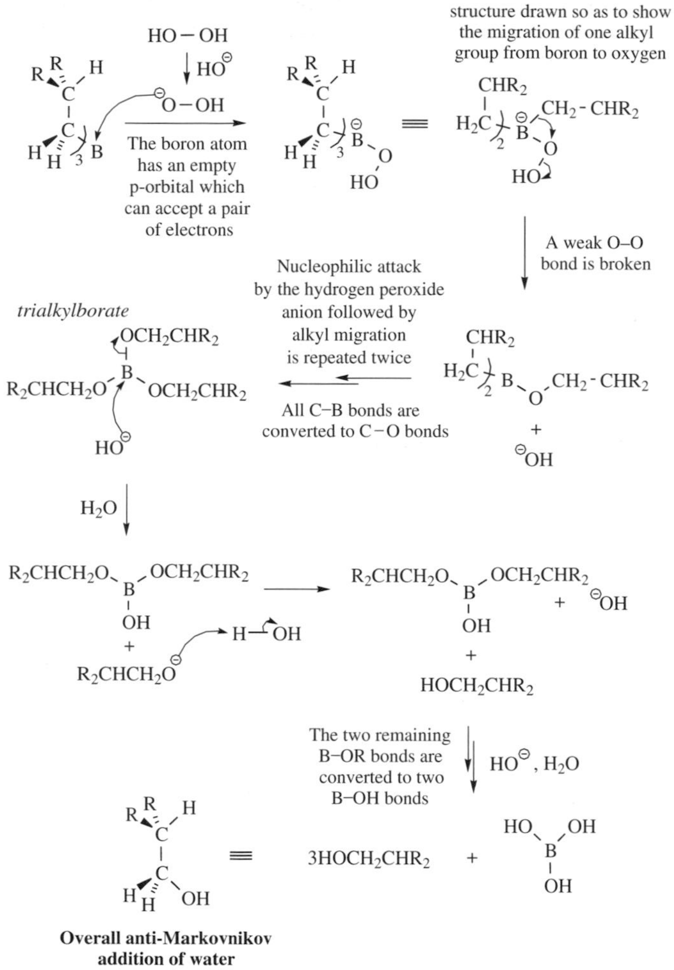

Overall anti-Markovnikov addition of water

The migration of the alkyl groups from boron to oxygen occurs with the retention of configuration.

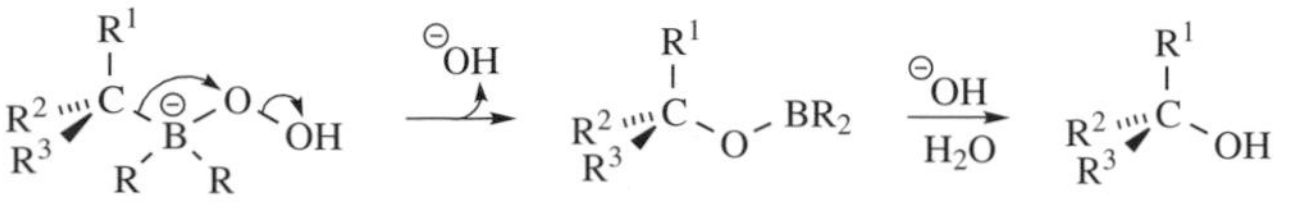

The boron group is replaced by an hydroxyl group with the same stereochemistry

6.2.2.6 Oxidation by peroxycarboxylic acids (RCO₃H) and hydrolysis of epoxides to give *anti-* dihydroxylation

The reaction of alkenes with peroxycarboxylic acids (or peracids) leads to the formation of epoxides in a concerted addition reaction. The peroxycarboxylic acid donates an oxygen atom to the double bond in a *syn-* addition reaction.

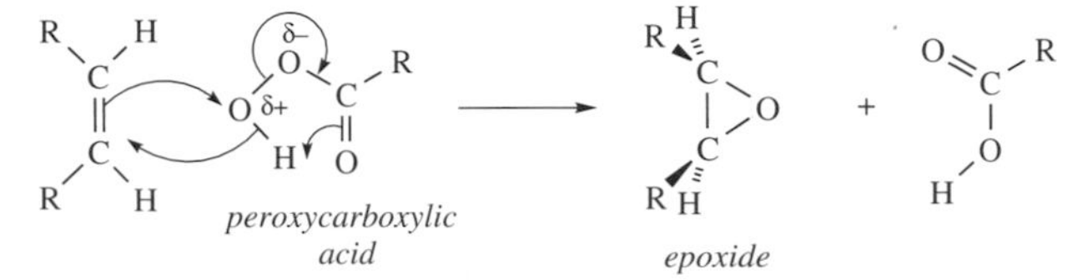

peroxycarboxylic acid

epoxide

The addition is stereospecific, so the *cis*-epoxide is formed from the *cis*-alkene.
A *trans*-alkene would give a *trans*-epoxide

Epoxides can be hydrolysed under acid- or base-catalysed conditions to form 1,2-diols. These reactions lead to the opening of the strained 3-membered (epoxide) ring. The formation and hydrolysis of an epoxide lead to the stereoselective *anti-* addition of two OH groups.

Base catalysis (aqueous hydroxide and heat)

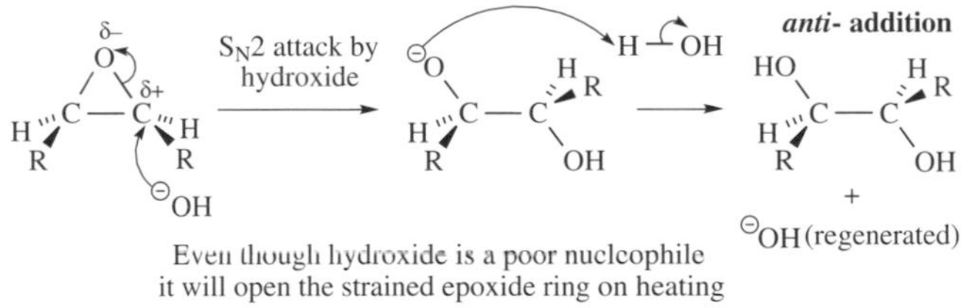

S_N2 attack by hydroxide

anti- addition

Even though hydroxide is a poor nucleophile it will open the strained epoxide ring on heating

Acid catalysis (aqueous acid)

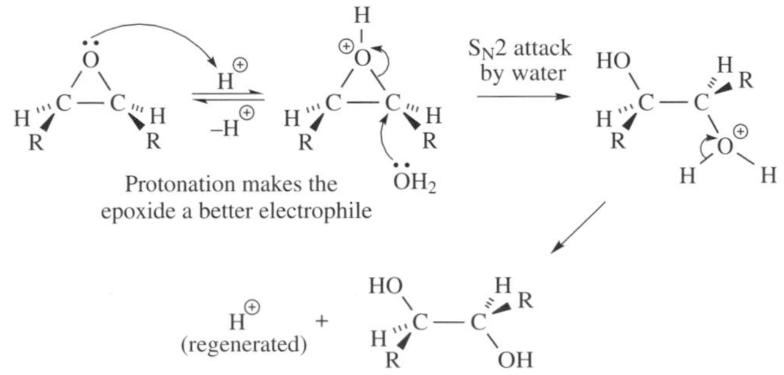

Protonation makes the epoxide a better electrophile

S_N2 attack by water

(regenerated)

6.2.2.7 Syn- dihydroxylation and oxidative cleavage of 1,2-diols to form carbonyls

The reaction of alkenes with potassium permanganate ($KMnO_4$) at low temperature, or osmium tetroxide (OsO_4), leads to the *syn-* addition of two OH groups (i.e. the two OH groups add to the same face of the double bond).

Potassium permanganate (low temperature)

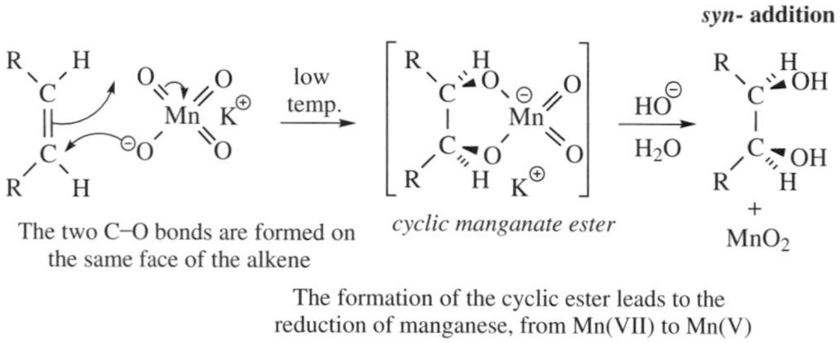

The two C–O bonds are formed on the same face of the alkene

cyclic manganate ester

The formation of the cyclic ester leads to the reduction of manganese, from Mn(VII) to Mn(V)

Osmium tetroxide

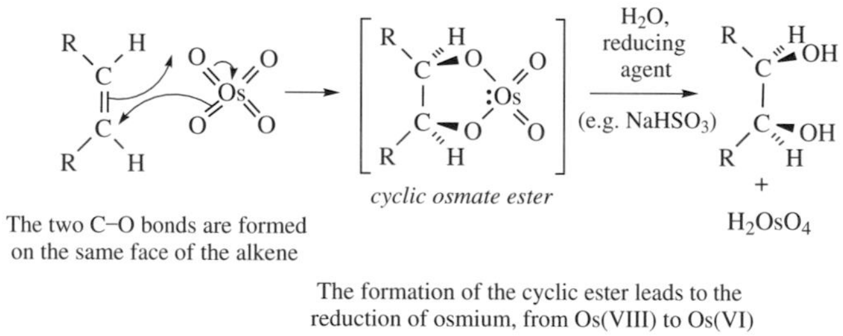

The two C–O bonds are formed on the same face of the alkene

cyclic osmate ester

The formation of the cyclic ester leads to the reduction of osmium, from Os(VIII) to Os(VI)

1,2-Diols can be oxidised further to form carbonyl compounds. This results in the cleavage of the C–C bond. When potassium permanganate is reacted with alkenes at room temperature or above, the intermediate 1,2-diol can be further oxidised to form carboxylic acids.

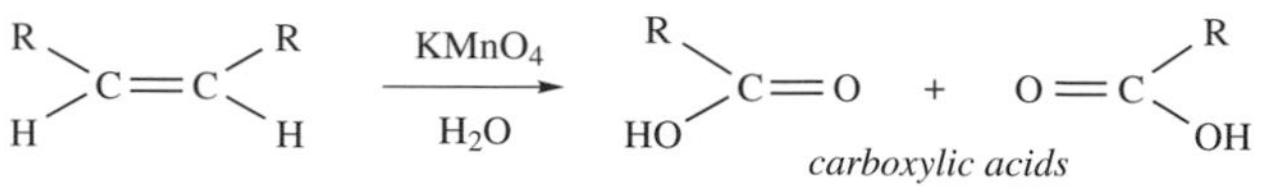

carboxylic acids

1,2-Diols can also be oxidised to aldehydes or ketones by reaction with periodic acid (HIO_4).

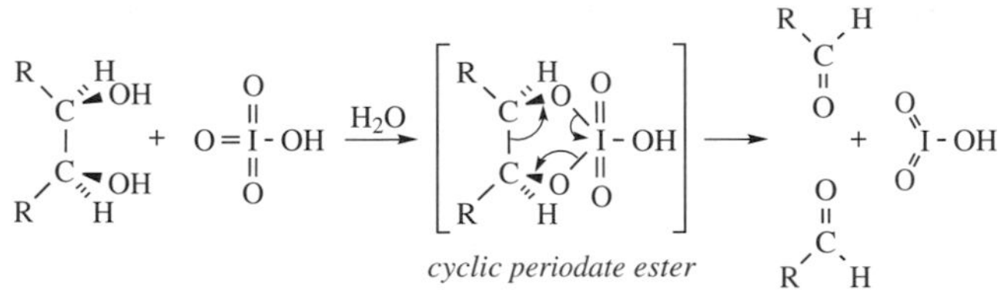

cyclic periodate ester

6.2.2.8 Oxidative cleavage by reaction with ozone (O_3)

The reaction of alkenes with ozone at low temperature produces an intermediate molozonide (or primary ozonide) which rapidly rearranges to form an ozonide. This leads to the cleavage of the C=C double bond.

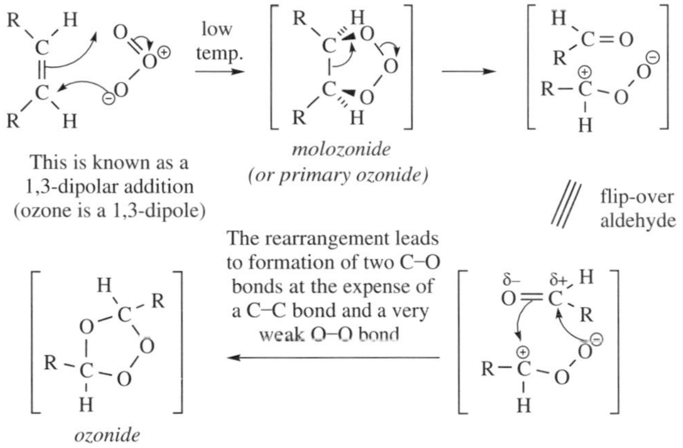

This is known as a
1,3-dipolar addition
(ozone is a 1,3-dipole)

molozonide
(or primary ozonide)

The rearrangement leads
to formation of two C–O
bonds at the expense of
a C–C bond and a very
weak O–O bond

flip-over
aldehyde

ozonide

The intermediate ozonide can then be reduced to aldehydes/ketones or be oxidised to carboxylic acids.

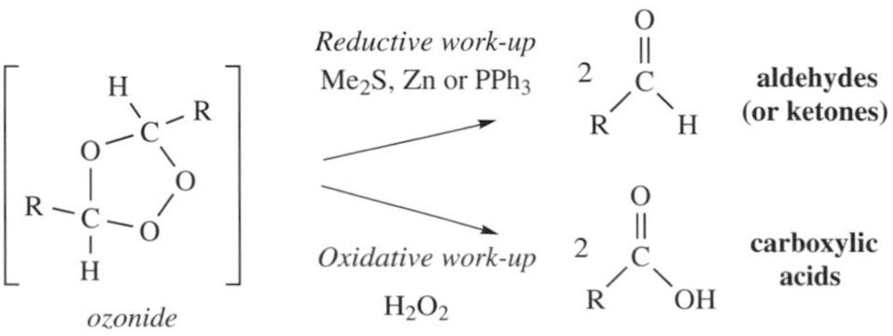

Reductive work-up
Me_2S, Zn or PPh_3

aldehydes
(or ketones)

Oxidative work-up
H_2O_2

carboxylic
acids

ozonide

6.2.2.9   Reaction with carbenes

A carbene is a neutral, highly reactive molecule containing a divalent carbon with only six valence electrons. They are usually formed in one of the two ways.

(1)   Reaction of trichloromethane (chloroform) and base. This is known as an *α-elimination* reaction, as two groups (i.e. H and Cl) are eliminated from the *same* atom.

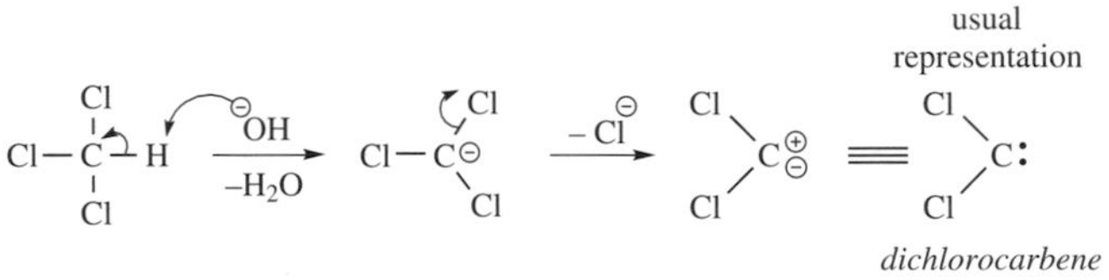

usual
representation

dichlorocarbene

(2)   Reaction of diiodomethane and a zinc–copper alloy (in the Simmons–Smith reaction).

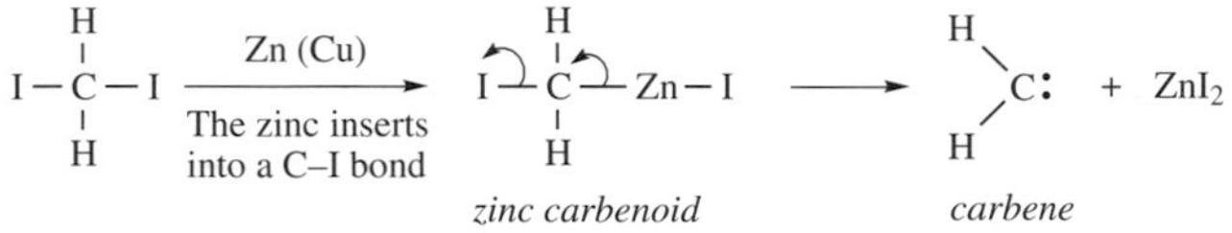

zinc carbenoid   *carbene*

As carbenes are electron deficient, they can act as powerful electrophiles. At the same time, they can also be thought of acting as nucleophiles, because they contain an unshared pair of electrons.

Carbenes will add to alkenes to form cyclopropanes. The mechanism of these insertion reactions depends on whether a singlet carbene (this contains a pair of reactive electrons in the same orbital) or a triplet carbene (this contains two unpaired and reactive electrons in different orbitals) adds to the double bond.

Singlet carbene (concerted mechanism – stereospecific syn- addition)

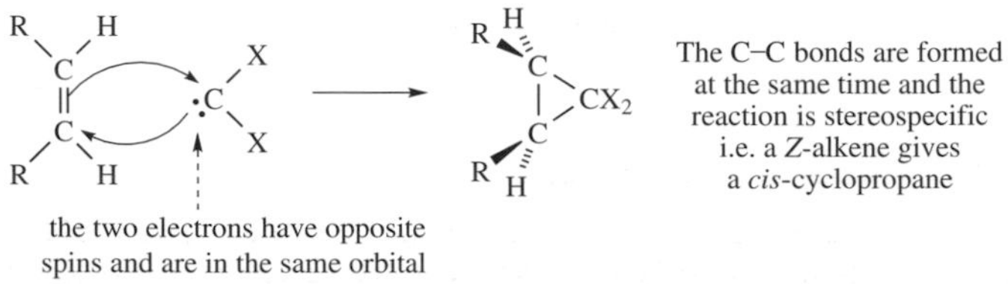

the two electrons have opposite
spins and are in the same orbital

The C–C bonds are formed
at the same time and the
reaction is stereospecific
i.e. a Z-alkene gives
a *cis*-cyclopropane

Triplet carbene (stepwise mechanism – non-stereospecific)

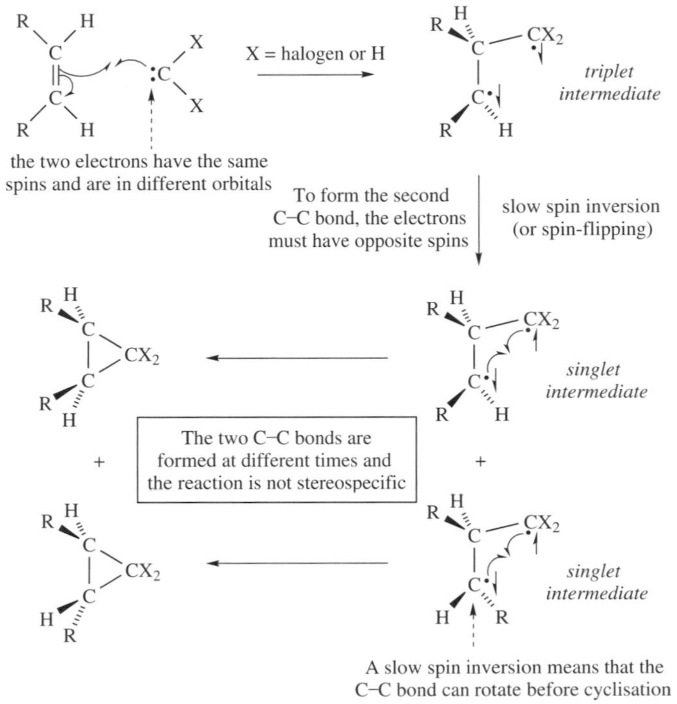

X = halogen or H

triplet intermediate

the two electrons have the same spins and are in different orbitals

To form the second C–C bond, the electrons must have opposite spins

slow spin inversion (or spin-flipping)

singlet intermediate

The two C–C bonds are formed at different times and the reaction is not stereospecific

singlet intermediate

A slow spin inversion means that the C–C bond can rotate before cyclisation

6.2.2.10   Addition of hydrogen (reduction)

The alkene double bond can be hydrogenated by hydrogen in the presence of a platinum or palladium catalyst. The hydrogenation occurs with *syn-* stereochemistry, as the two hydrogen atoms are added to the same face of the alkene on the surface of the catalyst.

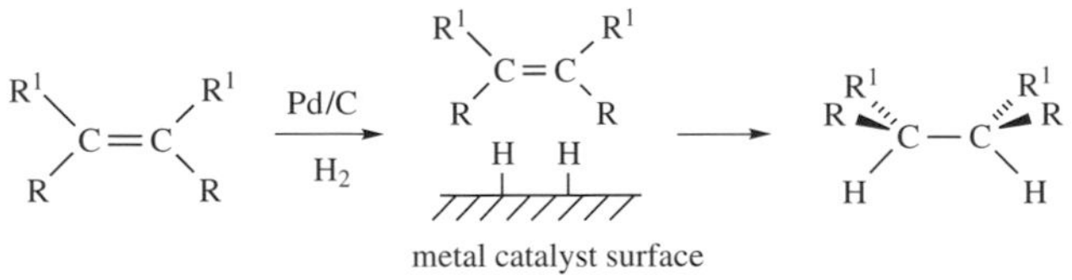

metal catalyst surface

6.2.2.11   Reaction with dienes

Alkenes bearing an electron-withdrawing substituent(s) can add to electron-rich conjugated dienes in the *Diels–Alder cycloaddition reaction*.

This is a concerted reaction leading to the formation of two new C–C bonds in one step. It is an important method for making rings, including cyclohexenes.

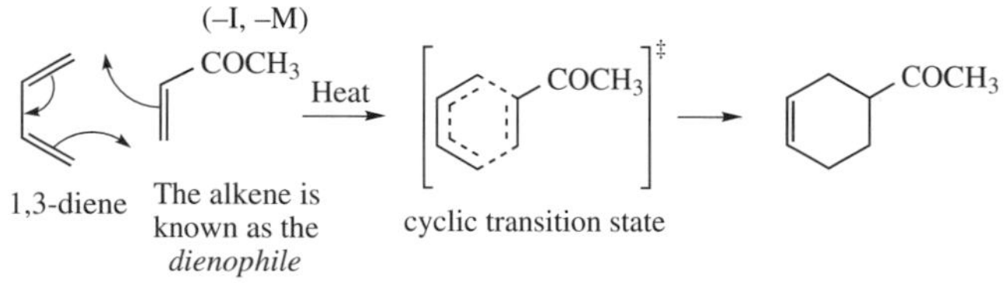

This is a *pericyclic* reaction, which involves a concerted redistribution of bonding electrons (i.e. a pericyclic reaction is a concerted reaction that involves a flow of electrons). The preference for electron-rich dienes and electron-deficient dienophiles arises from orbital interactions; this combination gives a good overlap of orbitals in the transition state.

The s-*cis* (or cisoid) conformation of the diene is required, and the reaction is stereospecific (i.e. only *syn*- addition). Groups that are *cis* in the (alkene) dienophile will therefore be *cis* in the product. Conversely, groups that are *trans* in the dienophile will be *trans* in the product.

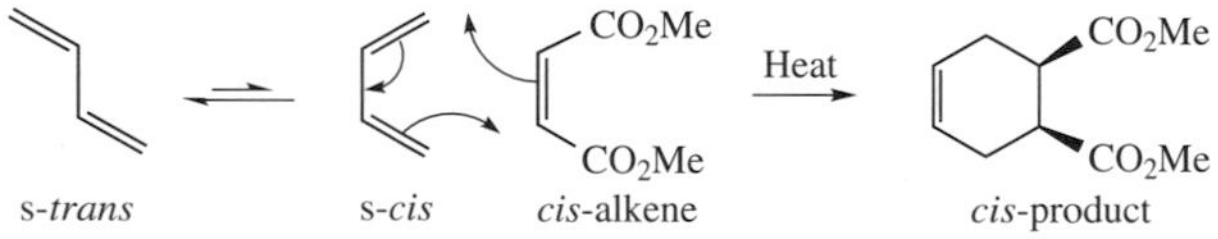

6.3 Alkynes

6.3.1 Preparation

Alkynes can be prepared from alkyl 1,2-dihalides (or vicinal dihalides) by the elimination of two molecules of HX using a strong base (e.g. NaNH$_2$).

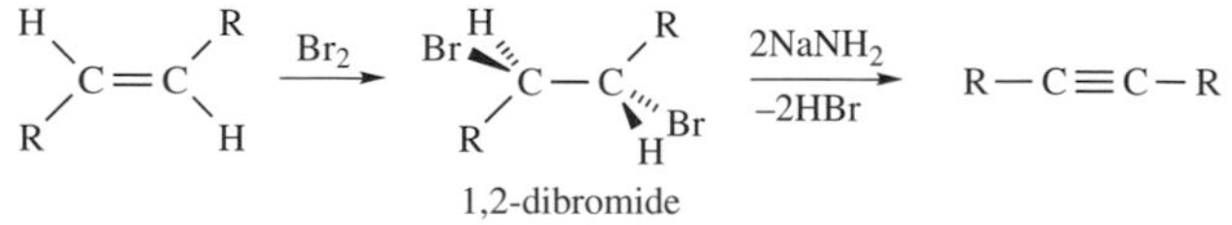

6.3.2 Reactions

Alkynes, like alkenes, act as nucleophiles and react with electrophiles in electrophilic addition reactions. The electrophiles generally add in the same way as they add to alkenes.

6.3.2.1 Addition of hydrogen halides (HX)

The addition of HX occurs so as to give the most substituted alkyl halide, following Markovnikov's rule (see Section 6.2.2.1). One or two equivalents of HX can be used to form vinyl halides or dihaloalkanes, respectively.

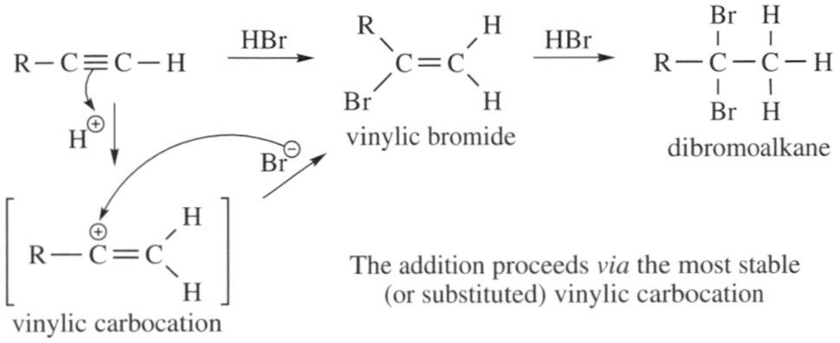

The addition proceeds *via* the most stable (or substituted) vinylic carbocation

Vinylic carbocations are generally less stable than alkyl carbocations, as there are fewer +I alkyl groups to stabilise the positive charge. As a consequence, alkynes (which give vinylic carbocations) generally react more slowly than alkenes (which give alkyl carbocations) in electrophilic addition reactions.

6.3.2.2 Addition of water (hydration): Markovnikov addition

Alkynes will react with water in the presence of a mercury(II) catalyst. Water adds in a Markovnikov addition to form an enol, which tautomerises to give a ketone (see Section 8.4.1).

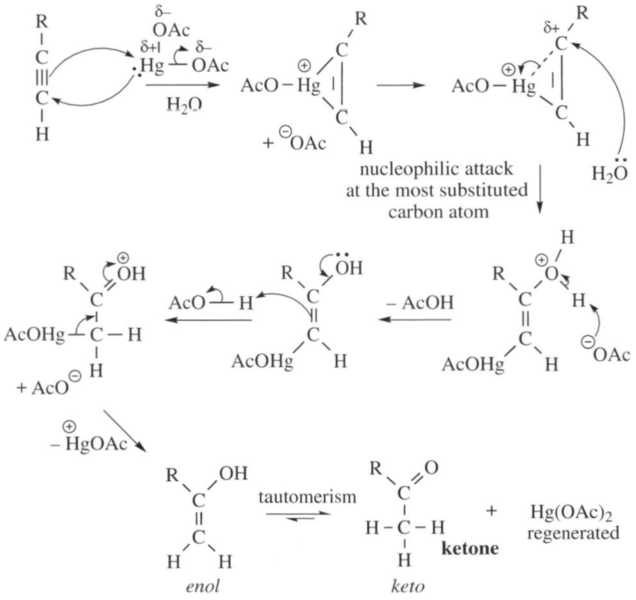

6.3.2.3 Addition of water (hydration): anti-Markovnikov addition

The hydroboration of alkynes using borane (BH_3) produces an intermediate vinylic borane, which then reacts with BH_3 in a second hydroboration; the two boron groups add to the less hindered end of the triple bond. However, when a bulky dialkylborane (HBR_2) is used, steric hindrance prevents a second hydroboration, and oxidation of the intermediate vinylic borane produces an enol, which tautomerises to an aldehyde. (For a related mechanism, see Section 6.2.2.5.)

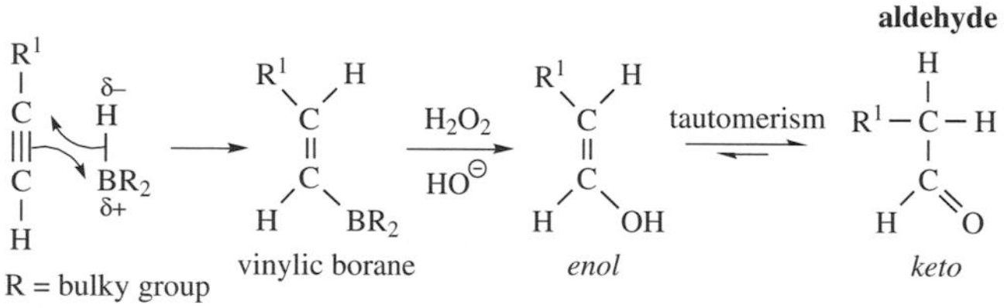

This results in the anti-Markovnikov addition of water to the triple bond.

6.3.2.4 Addition of hydrogen (reduction)

Alkynes can be stereoselectively reduced to give Z- or E-alkenes using H_2/Lindlar catalyst or Na/NH_3 (at low temperature), respectively.

Z-Alkene

The metal catalyst ensures that the two hydrogen atoms add to the same face of the alkene in a *syn*- addition.

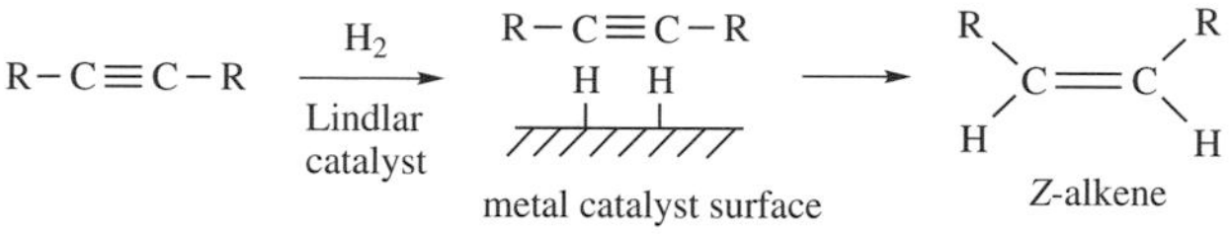

The Lindlar catalyst [$Pd/CaCO_3/PbO$] is a poisoned palladium catalyst, which ensures that the reduction stops at the alkene (and does not go on to give an alkane). Reduction to the alkane requires a Pd/C catalyst (see Section 6.2.2.10).

E-Alkene

The two hydrogen atoms add to the opposite faces of the alkene (i.e. *anti*-addition) using sodium in liquid ammonia. This 'dissolving metal reduction' produces a solvated electron, which adds to the alkyne to produce a radical anion (bearing a negative charge and an unpaired electron). The bulky R groups in the radical anion, vinyl radical and subsequently in the vinyl anion lie as far apart as possible.

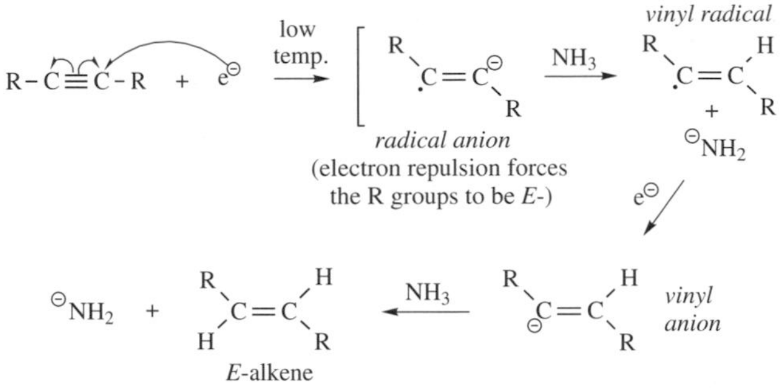

6.3.2.5 Deprotonation: formation of alkynyl anions

Terminal alkynes contain an acidic proton, which can be deprotonated by sodium amide ($NaNH_2$). The negative charge of the alkynyl (or acetylide) anion resides in an sp-orbital. This is more stable than a vinyl anion (produced on deprotonation of an alkene), because these are sp^2 hybridised. The greater the 's' character, the more closely the anion is held to the positively charged nucleus (which stabilises it).

The alkynyl anion can act as a nucleophile. Reaction with primary alkyl halides will lead to alkylation (by an S_N2 mechanism) and the introduction of an alkyl group on the terminal carbon atom of the alkyne.

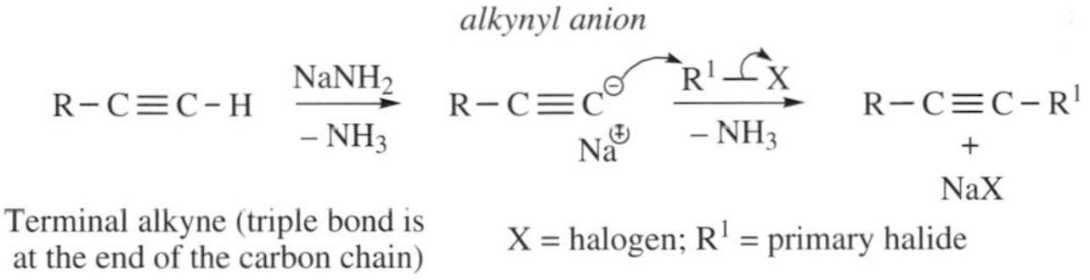

Problems

(1) Give mechanisms for the reaction of HBr with pent-1-ene in (a) the presence of a peroxide (ROOR) and (b) in the absence of a peroxide. Explain how your mechanisms account for the regioselectivity of the reactions.

(2) Give a mechanism for the reaction of Br_2 with (Z)-hex-3-ene, and explain how the mechanism accounts for the stereospecificity of the reaction.

(3) Suggest suitable reagents to accomplish the following trans-formations. Note that the *relative* rather than the *absolute stereo-chemistry* of the products is shown (i.e. even though just one

enantiomer of the product is drawn, the product is racemic, because
the alkene starting material is achiral).

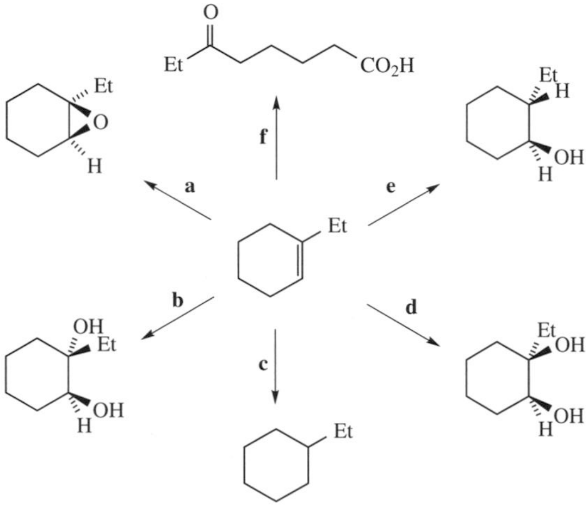

(4) Suggest syntheses of compounds **A, B** and **C** starting from the
     alkyne $CH_3CH_2CH_2C{\equiv}CH$ (pent-1-yne).

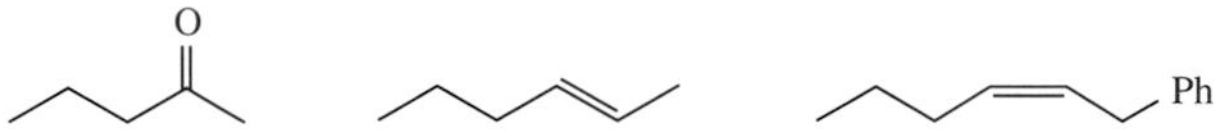

A   **B**   **C**

(5) Give the diene and dienophile that would react in a Diels–Alder
     reaction to give each of the products **D–F**.

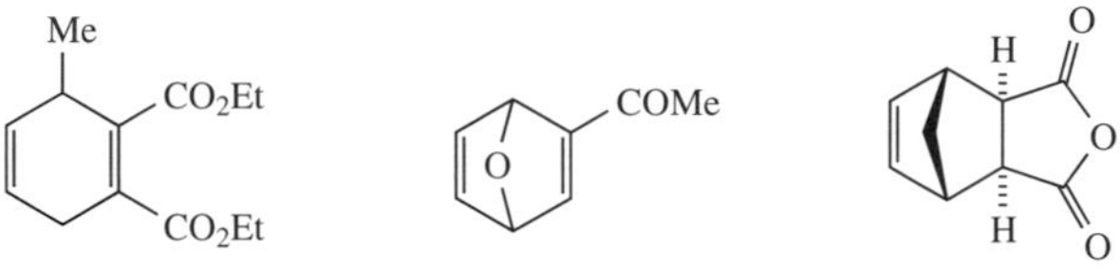

D   **E**   **F**

(6) Explain why the reaction of buta-1,3-diene ($H_2C{=}CH{-}CH{=}CH_2$)
     with one equivalent of HCl produces both 3-chlorobut-1-ene
     (~80%) and 1-chlorobut-2-ene (~20%).

7. BENZENES

Key point. Benzene is an aromatic compound because the six π-electrons are delocalised over the planar 6-membered ring. The delocalisation of electrons results in an increase in stability, and benzene is therefore less reactive than alkenes or alkynes. Benzene generally undergoes *electrophilic substitution reactions*, in which a hydrogen atom is substituted for an electrophile. The electron-rich benzene ring attacks an electrophile to form a carbocation, which rapidly loses a proton so as to regenerate the aromatic ring. Electron-donating (+I, +M) substituents (on the benzene ring) make the ring more reactive towards further electrophilic substitution and direct the electrophile to the *ortho-/para-* positions. In contrast, electron-withdrawing (−I, −M) substituents (on the benzene ring) make the ring less reactive towards further electrophilic substitution and direct the electrophile to the *meta-* position.

7.1 Structure

- Benzene (C_6H_6) has six sp^2 carbon atoms and is cyclic, conjugated and planar. It is symmetrical and all C−C−C bond angles are 120°. The six C−C bonds are all 1.39 Å long, which is in between the normal values for a C−C and C=C bond.
- Benzene has six π-electrons, which are delocalised around the ring, and a circle in the centre of a 6-membered ring can represent this. However, it is generally shown as a ring with three C=C bonds, because it is easier to draw reaction mechanisms using this representation.

The six p-orbitals overlap to give 'rings' of electron density above and below the six-carbon ring

Circle denotes delocalisation of electrons

Shown as a single resonance form

As benzene has six π-electrons, it obeys Hückel's rule and is *aromatic*. Hückel's rule states that only cyclic planar molecules with $4n + 2$ π-electrons can be aromatic: for benzene $n = 1$. (Systems with $4n$ π-electrons are described as *anti-aromatic*.)

Aromatic compounds can be monocyclic or polycyclic, neutral or charged. Atoms other than carbon can also be part of the ring, and for pyridine, the lone pair of electrons on nitrogen is not part of the π-electron system (see Section 1.7.5).

Examples (each double bond represents two π-electrons)

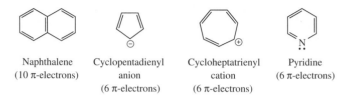

Naphthalene	Cyclopentadienyl	Cycloheptatrienyl	Pyridine
(10 π-electrons)	anion	cation	(6 π-electrons)
	(6 π-electrons)	(6 π-electrons)	

For benzene, the molecular orbital theory states that the six p-orbitals combine to give six molecular orbitals. The three lower-energy molecular orbitals are bonding molecular orbitals, and these are completely filled by the six electrons (which are spin-paired). There are no electrons in the (higher-energy) antibonding orbitals, and hence benzene has a closed bonding shell of delocalised π-electrons.

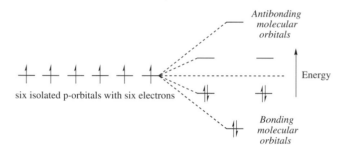

The delocalisation of electrons in aromatic compounds gives rise to characteristic chemical shift values in the ^{1}H NMR spectra (see Section 10.5.1).

7.2 Reactions

The delocalisation of electrons means that benzene is unusually stable. It has a heat of hydrogenation which is approximately $150 \, \text{kJ mol}^{-1}$ less than that which would be expected for a cyclic conjugated triene. This is known as the *resonance energy*.

The unusual stability of benzene (and other aromatic molecules) means that it undergoes *substitution* rather than addition reactions (cf. alkenes/alkynes). This is because substitution reactions lead to the formation of products which retain the stable aromatic ring.

As benzene contains six π-electrons, it can act as a nucleophile and react with electrophiles in *electrophilic substitution reactions*. These reactions involve the substitution of a hydrogen atom on the benzene ring for an electrophile. The initial electrophilic attack on the benzene ring leads to the formation of a positively charged intermediate (known as a *Wheland intermediate*), which is readily deprotonated.

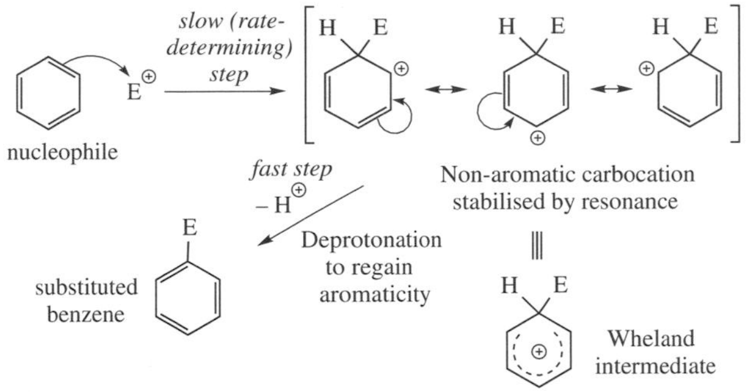

7.2.1 Halogenation

The reaction of benzene with bromine or chlorine in the presence of a Lewis acid catalyst (such as $FeBr_3$, $FeCl_3$ or $AlCl_3$) leads to the formation of bromobenzene or chlorobenzene, respectively. The Lewis acid, which does not have a full outer electron shell, can form a complex with bromine or chlorine. This polarises the halogen–halogen bond (making the halogen more electrophilic), and attack occurs at the positive end of the complex.

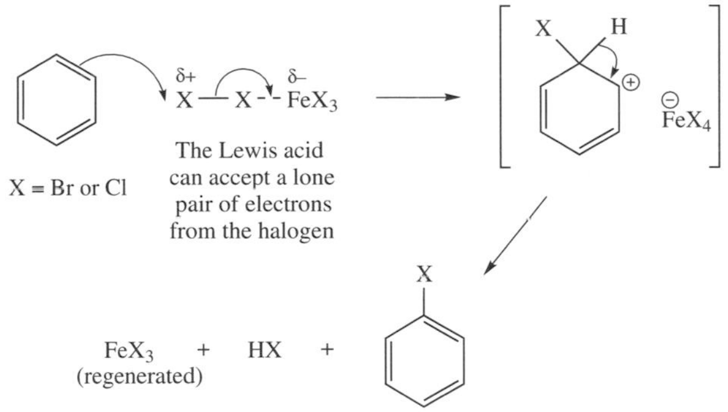

No reaction occurs in the absence of a Lewis acid. This contrasts with the halogenation of alkenes/alkynes, which does not require activation by a Lewis acid. This is because alkenes/alkynes are not aromatic, and hence they are more reactive nucleophiles than benzene.

7.2.2 Nitration

Benzene can be nitrated using a mixture of concentrated nitric (HNO_3) and sulfuric (H_2SO_4) acid. These acids react to form an intermediate nitronium ion, which acts as the electrophile.

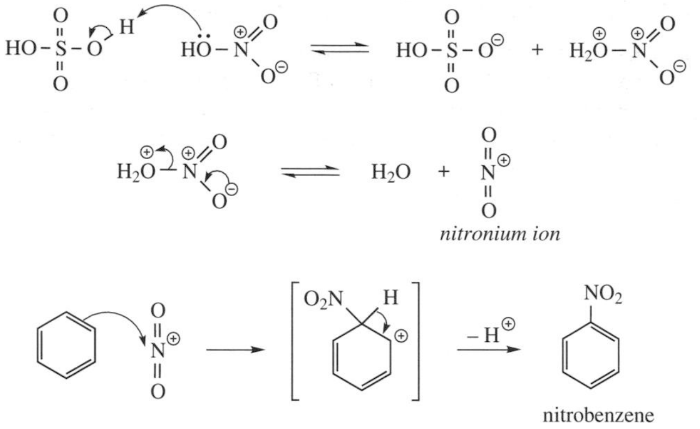

nitronium ion

nitrobenzene

7.2.3 Sulfonation

The reaction of benzene with fuming sulfuric acid or oleum (a mixture of sulfuric acid and sulfur trioxide) leads to the formation of benzenesulfonic acid. The sulfonation is reversible, and this makes it a useful tool in organic synthesis for blocking certain positions on a benzene ring (see Section 7.8). Sulfonation is favoured using strong sulfuric acid, but desulfonation is favoured in hot, dilute aqueous acid.

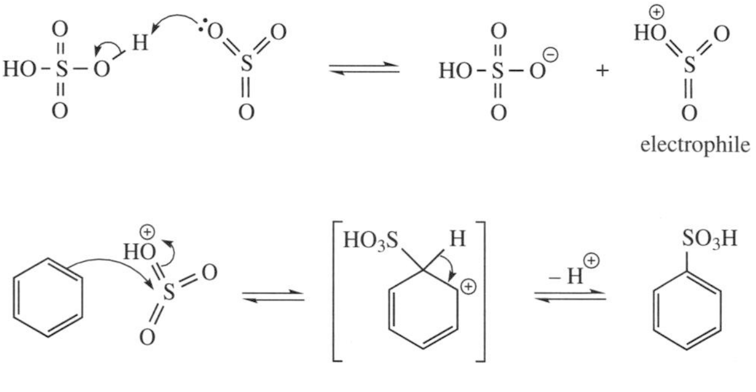

electrophile

Sulfur trioxide (rather than protonated sulfur trioxide) could also act as the electrophile in these reactions.

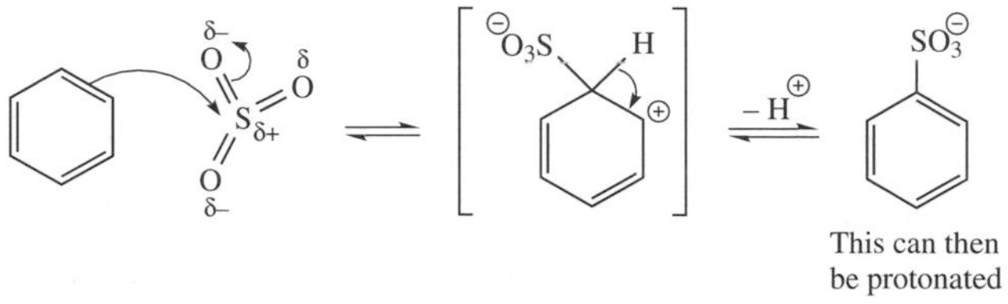

This can then
be protonated

7.2.4   Alkylation: the Friedel–Crafts alkylation

The reaction of benzene with alkyl bromides or chlorides in the presence of a Lewis acid catalyst (such as $FeBr_3$, $FeCl_3$ or $AlCl_3$) leads to the formation of alkylbenzenes. (Aryl or vinyl halides (ArX, $R_2C=CHX$) do not react.) The Lewis acid increases the electrophilicity of the carbon atom attached to the halogen. This can lead to the formation of a carbocation, which then reacts with the electron-rich benzene ring.

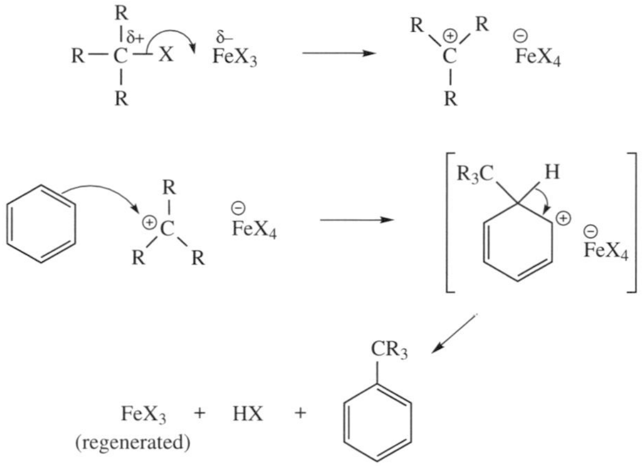

Intermediate primary or secondary carbocations may rearrange to give more stable secondary or tertiary carbocations, before the carbon–carbon bond formation can take place (see Section 6.2.2.1). In general, the higher the reaction temperature, the greater the amount of rearranged product.

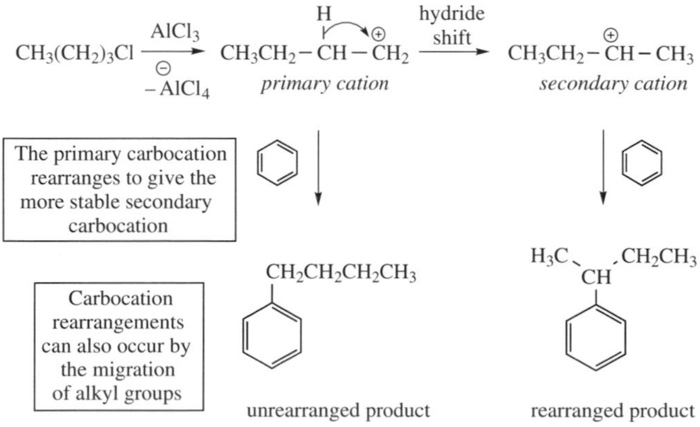

7.2.5   Acylation: the Friedel–Crafts acylation

The reaction of benzene with acid chlorides in the presence of a Lewis acid catalyst (such as $FeCl_3$ or $AlCl_3$) leads to the formation of acyl-benzenes. The Lewis acid increases the electrophilicity of the carbon atom attached to the chlorine. This leads to the formation of an acylium ion, which reacts with the electron-rich benzene ring.

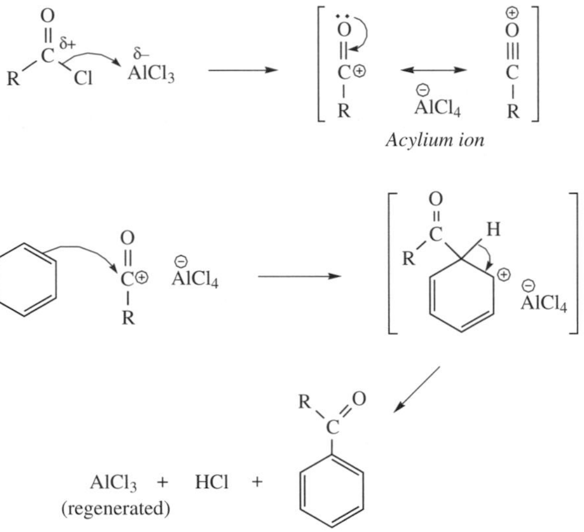

Acylium ion

$$AlCl_3 \; + \; HCl \; +$$
(regenerated)

Unlike carbocations, the intermediate acylium ion does not rearrange and is attacked by the benzene ring to give exclusively the unrearranged product.

The Gattermann–Koch reaction (which uses CO, HCl and $AlCl_3$) can be used to produce an aromatic aldehyde, rather than a ketone.

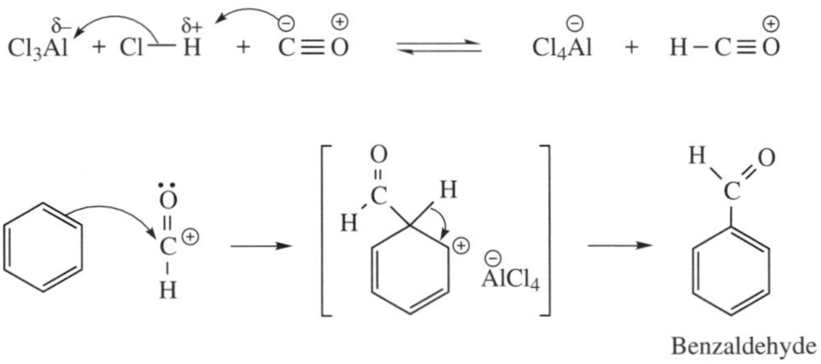

Benzaldehyde

7.3   Reactivity of substituted benzenes

The introduction of substituents on the benzene ring affects both the reactivity of the benzene ring and also the regiochemistry of the reaction (i.e. the position in which the new group is introduced on the benzene ring).

7.3.1   Reactivity of benzene rings: activating and deactivating substituents

- Substituents that donate electron density towards the benzene ring are known as the *activating groups*. These groups, which have positive inductive (+I) and/or mesomeric effects (+M), make the substituted benzene ring more reactive to electrophilic substitution than benzene itself. This is because the activating group can help to stabilise the carbocation intermediate (by donating electrons).

- Substituents that withdraw electron density away from the benzene ring are known as the *deactivating groups*. These groups, which have negative inductive (−I) and/or mesomeric effects (−M), make the substituted benzene ring less reactive to electrophilic substitution than benzene itself. This is because the deactivating group can destabilise the carbocation intermediate (by withdrawing electrons).

EDG (+I, +M)   EDG

More reactive than benzene   EDG stabilises the carbocation

E⊕

EDG = electron-donating group

EWG (−I, −M)   EWG

Less reactive than benzene   EWG destabilises the carbocation

E⊕

EWG = electron-withdrawing group

- The greater the +M and/or +I effect, the more activating the group and the more reactive the benzene ring is to electrophilic attack. Note that positive mesomeric effects are generally stronger than positive inductive effects (see Section 1.6).

- The greater the −M and/or −I effect, the more deactivating the group and the less reactive the benzene ring is to electrophilic attack.

Activating groups		Deactivating groups	
NHR, NH$_2$ (+M, −I)	*Most reactive*	Cl, Br, I (+M, −I)	
OR, OH (+M, −I)	↑	CHO, COR (−M, −I)	
NHCOCH$_3$ (+M, −I)		CO$_2$H, CO$_2$R (−M, −I)	
Aryl (Ar) (+M, +I)		SO$_3$H (−M, −I)	
Alkyl (R) (+I)		NO$_2$ (−M, −I)	*Least reactive*

7.3.2   Orientation of reactions

Electrophilic substitution can occur at the *ortho-* (2-/6-), *meta-* (3-/5-) or *para-* (4-) position of the benzene ring (see Section 2.4). The inductive and/or mesomeric effects of the existing substituent determine which position the new substituent is introduced on the ring. On introducing a new substituent, the formation of the carbocation intermediate is the rate-determining step. If the existing substituent can stabilise the carbocation, then this will lower the activation energy, leading to attack at this position.

Substituents can be classified as: (i) *ortho-/para-*directing activators; (ii) *ortho-/para-*directing deactivators; or (iii) *meta-*directing deactivators.

ortho-/para- activators	*ortho-/para-* deactivators	*meta-* deactivators
NHR, NH$_2$	Cl, Br, I	CHO, COR
OR, OH		CO$_2$H, CO$_2$R
NHCOCH$_3$		SO$_3$H
Aryl (Ar)		NO$_2$
Alkyl (R)		

The relative ease of attack (by an electrophile) at the *ortho-/meta-/para-*positions can be determined experimentally from *partial rate factors.* These compare the rate of attack at one position in the monosubstituted benzene with the rate of attack at one position in benzene. The higher the partial rate factor at a given position, the faster the rate of electrophilic substitution.

7.3.2.1   *Ortho-/para-*directing activators

Electron-donating +I and/or +M groups (EDGs), which make the ring more nucleophilic than benzene, will stabilise the intermediate carbocation most effectively when new substituents are introduced at the *ortho-* or *para-* position. For the *meta-* isomer, the positive charge in the carbocation intermediate does not reside adjacent to the EDG in any of the resonance forms.

EDG = electron-donating group

The EDG can stabilise both of these cations more effectively than the cation derived from *meta-* attack

An additional resonance structure can be drawn for intermediate carbocations bearing the +M but not +I groups (at the 2- or 4-position).

Examples

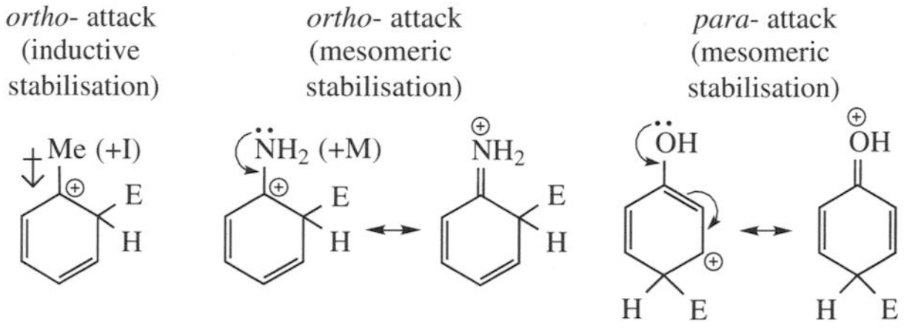

ortho- attack (inductive stabilisation)	ortho- attack (mesomeric stabilisation)	para- attack (mesomeric stabilisation)

7.3.2.2 *Ortho-/para*-directing deactivators

Halogen groups are unique in that they direct *ortho-/para-* and yet they deactivate the benzene ring. The +M effect of Cl, Br and I is weak because these atoms are all larger than carbon. This means that the orbitals containing the non-bonding pairs of electrons (e.g. 3p-orbitals for Cl) do not overlap well with the 2p-orbital of carbon. (This is a general phenomenon, and mesomeric effects are not transmitted well between atoms of different rows of the periodic table.) The weak +M effect does, however, ensure that the halogens are *ortho-/para-* directing, but the strong −I effect (which deactivates the ring) is more significant in terms of the reactivity of halobenzenes.

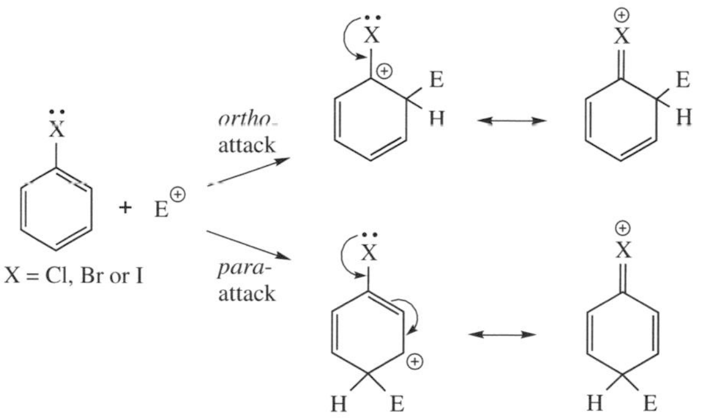

7.3.2.3 *Meta*-directing deactivators

Electron-withdrawing −I/−M groups (EWGs), which make the ring less nucleophilic than benzene, will deactivate the *meta-* position less than the *ortho-/para-* positions. The carbocation produced from attack at the *meta-* position will therefore be the most stable because this does not reside adjacent to the EWG in any of the resonance forms.

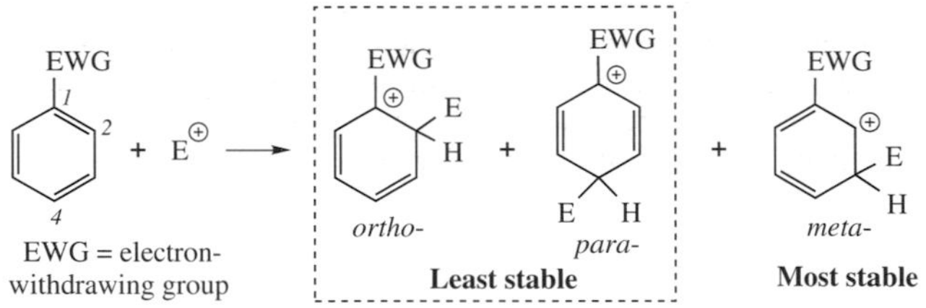

EWG = electron-
withdrawing group

ortho-   para-
Least stable

meta-
Most stable

The EWG can destabilise both of these cations more
effectively than the cation derived from *meta-* attack

Examples

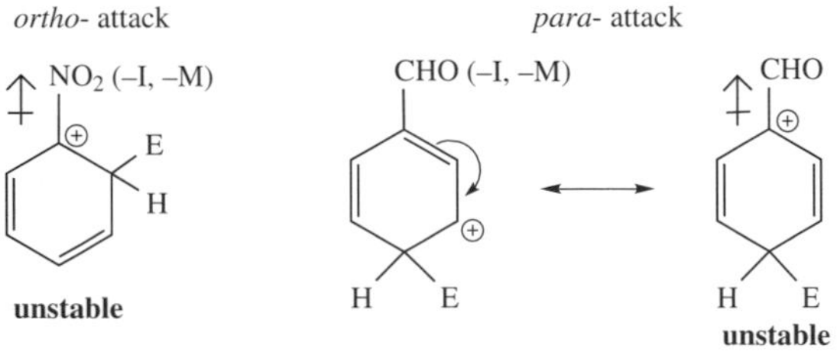

ortho- attack

NO$_2$ (–I, –M)

unstable

para- attack

CHO (–I, –M)

CHO

unstable

7.3.2.4 Steric effects versus electronic effects

With activating groups, we would expect the ratio of attack at the
ortho- and *para-* positions to be 2:1 (as there are two *ortho-* positions
to one *para-* position on the ring). However, attack at the *ortho-* position
is often less than this because of *steric hindrance*. The size of the group
on the benzene ring strongly influences the substitution at the adjacent
ortho- position. In general, the larger the size of the group on the ring,
the greater the proportion of *para-* substitution.

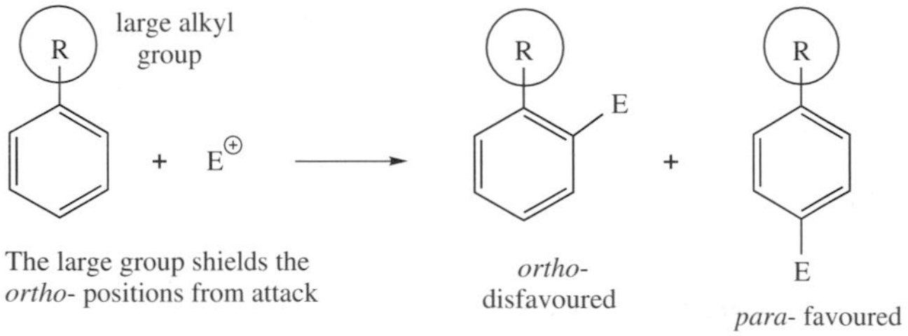

The large group shields the
ortho- positions from attack

ortho-
disfavoured

para- favoured

7.4 Nucleophilic aromatic substitution (the S$_N$Ar mechanism)

The reaction of a nucleophile with a benzene ring is very rare. This can
only occur when strongly electron-withdrawing substituents (e.g. NO$_2$)
are at the *ortho-* and/or *para-* positions of aryl halides. The reaction

proceeds by an addition–elimination mechanism, and the EWGs can stabilise the intermediate carbanion (called a Meisenheimer complex) by resonance. It should be noted that this is *not* an S_N2 reaction, as S_N2 reactions cannot take place at an sp^2 carbon atom (because the nucleophile cannot approach the C–X bond at an angle of 180°).

The order of reactivity of *aryl* halides is Ar–F >> Ar–Cl ~ Ar–Br ~ Ar–I, which is the reverse order for S_N2/S_N1 reactions of *alkyl* halides (see Section 5.3.1). As fluorine is the most electronegative halogen atom, it strongly polarises the C–F bond. This makes the carbon δ+, and hence it is susceptible to nucleophilic attack. The fluorine atom is also small, and hence the incoming nucleophile can easily approach the adjacent carbon atom (as there is little steric hindrance).

Aryl diazonium compounds (Ar–N_2^+) can also undergo substitution reactions in the presence of nucleophiles. However, the reactions appear to involve radical or ionic (S_N1-like) mechanisms. An S_N1 mechanism requires the formation of an aryl cation (Ar+), which is highly unstable because the positive charge cannot be delocalised around the ring. It appears, however, that these highly reactive cations can be formed (and trapped by various nucleophiles) because gaseous nitrogen (N_2) is an excellent leaving group (see Section 7.6).

7.5 The formation of benzyne

The reaction of halobenzenes with a strong base (e.g. $NaNH_2$ in NH_3) can lead to the elimination of HX and the formation of benzyne (in a synperiplanar elimination reaction). The new π-bond that is formed (by the overlap of two sp^2-orbitals outside the ring) is very weak, and benzyne can react with nucleophiles or take part (as the dienophile) in Diels–Alder reactions (see Section 6.2.2.11). As benzyne is extremely

unstable, it undergoes nucleophilic attack by $^-NH_2$ to give a new anion
(this is not typical of ordinary alkynes).

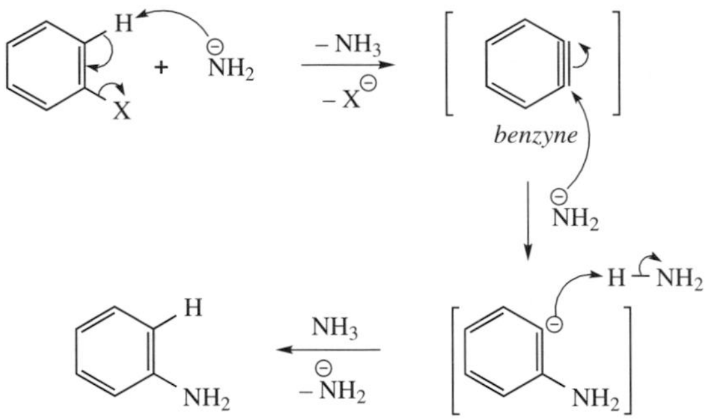

7.6   Transformation of side chains

Functional groups on a benzene ring can generally be converted to
other functional groups without reacting with the stable aromatic ring.

(1)   *Oxidation of alkyl side chains.*   Strong oxidising agents (e.g. $KMnO_4$)
      can convert alkyl side chains to carboxylic acids.

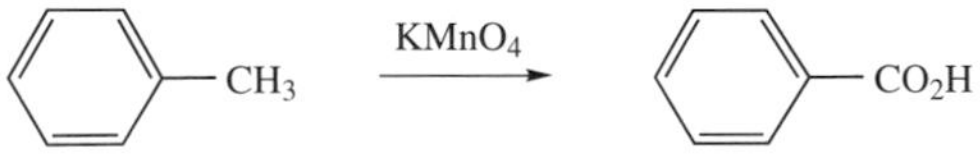

(2)   *Bromination of alkyl side chains.*   N-Bromosuccinimide (NBS) can
      brominate at the benzylic position *via* a radical chain mechanism
      (i.e. at the carbon atom attached to the benzene ring). It should be
      noted that the intermediate benzylic radical is stabilised by reso-
      nance (i.e. the radical can interact with the π-electrons of the
      benzene ring).

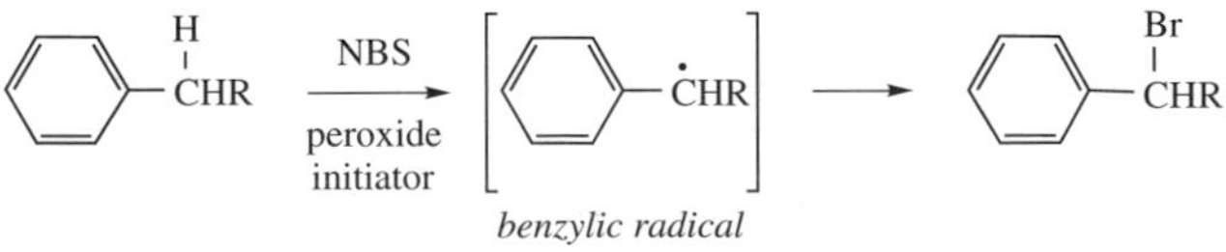

(3)   *Reduction of nitro groups.*   Catalytic hydrogenation or reaction
      with iron or tin in acid can achieve this.

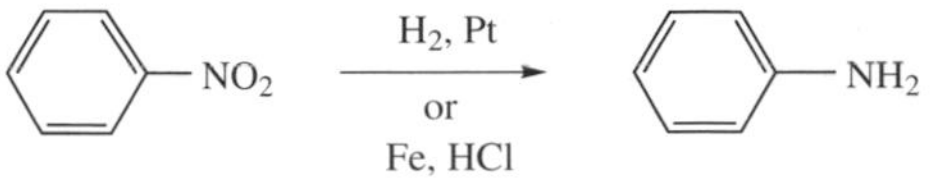

(4) *Reduction of ketone groups.* Catalytic hydrogenation or reaction with zinc amalgam in strong acid (in the *Clemmensen reduction*) can achieve this.

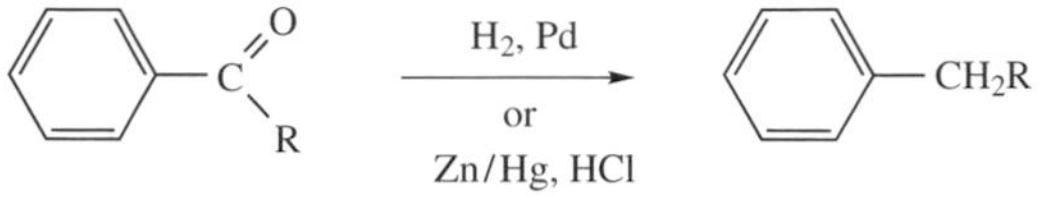

(5) *The preparation and reactions of diazonium salts.* The reaction of aromatic amines with the nitrosonium ion (^+NO), generated from nitrous acid (HNO_2), yields aryl diazonium salts, which can be converted (on loss of nitrogen) to a range of functional groups.

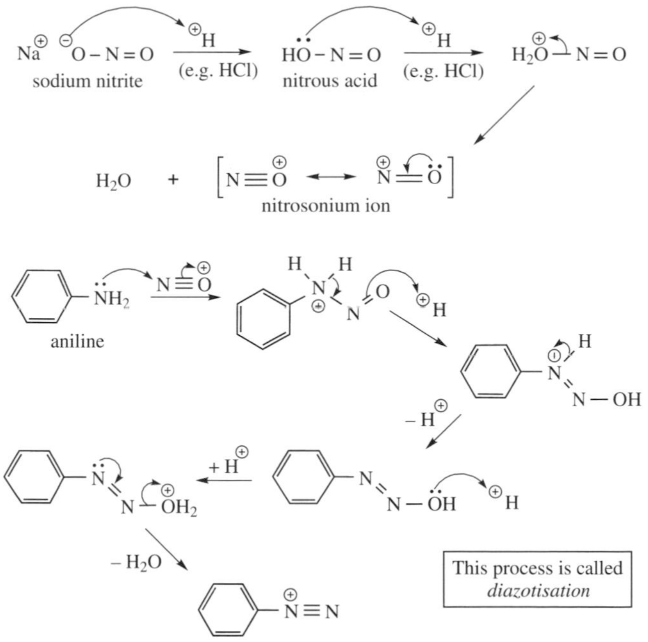

aryl diazonium ion

On heating, the aryl diazonium ion can produce nitrogen gas (an excellent leaving group) and a very unstable aryl cation in an S_N1 mechanism, which can react with nucleophiles.

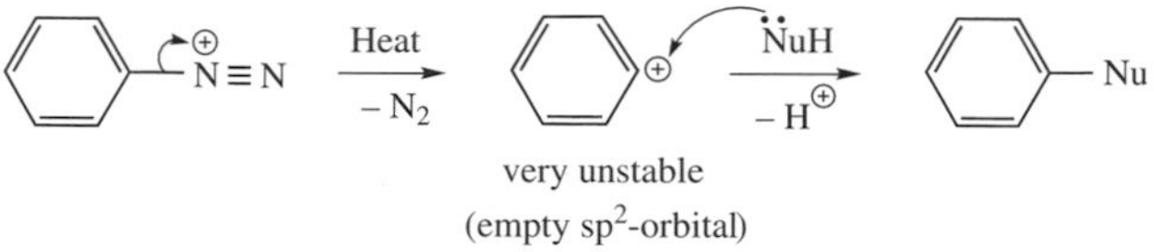

Heat
– N_2
NuH
– $H^{\oplus}$
Nu
very unstable
(empty sp^2-orbital)

Functional group transformations

Br
Cl
CN
Cu(I)Br
Cu(I)Cl
Cu(I)CN
HBF_4, heat
KI
F
I
H_2O, heat
H_3PO_2
OH
H

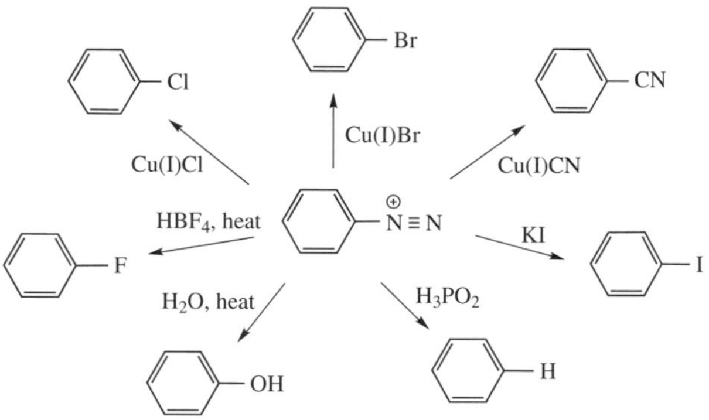

The reactions involving copper(I) salts are known as the *Sandmeyer reactions*. These reactions proceed *via* radical, rather than ionic, mechanisms.

Aryl diazonium salts can also undergo *coupling reactions* with phenol or aromatic amines, which possess nucleophilic OH or NH_2 groups, respectively. This electrophilic substitution reaction (with the diazonium salt as the electrophile) produces highly coloured azo compounds.

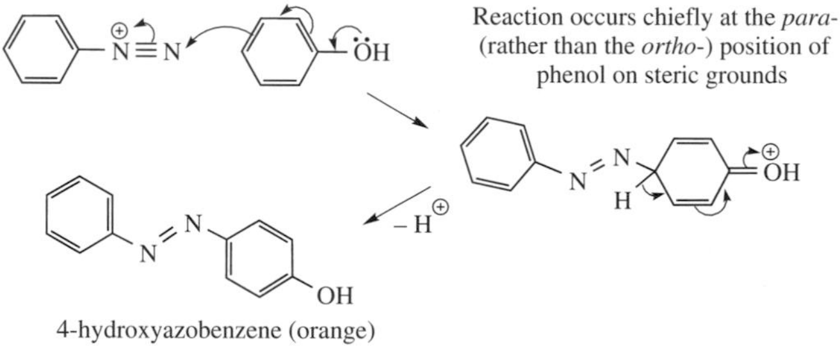

Reaction occurs chiefly at the *para*-
(rather than the *ortho*-) position of
phenol on steric grounds
– $H^{\oplus}$
OH
4-hydroxyazobenzene (orange)

7.7 Reduction of the benzene ring

Harsh reaction conditions are required to reduce the aromatic benzene ring. This can be achieved by catalytic hydrogenation (using high temperatures or pressures and very active catalysts) or alkali metals in liquid ammonia/ethanol (in a *Birch reduction*).

Catalytic hydrogenation

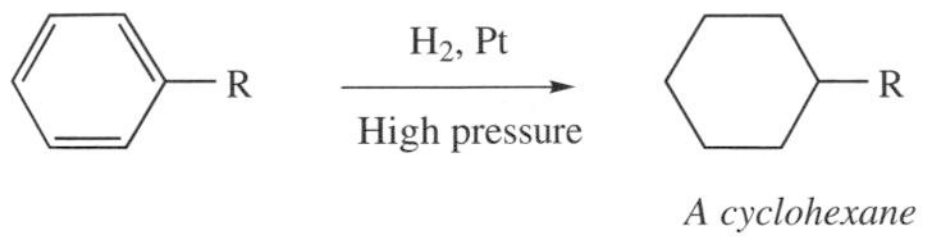

A cyclohexane

Birch reduction. Sodium or lithium metal (in liquid ammonia) can donate an electron to the benzene ring to form a radical anion. On protonation (by ethanol) and further reduction/protonation, this produces 1,4-cyclohexadiene.

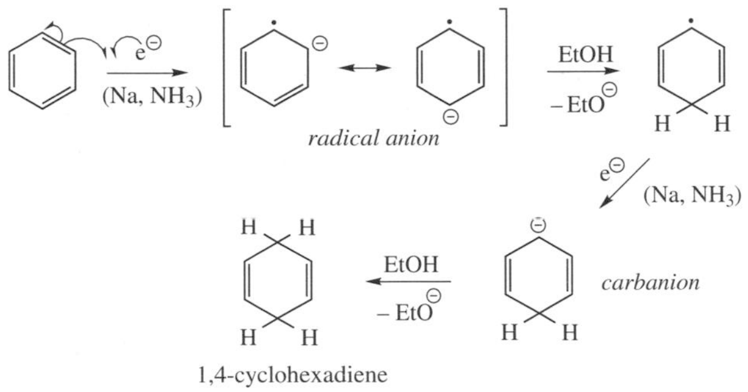

7.8 The synthesis of substituted benzenes

The following points should be borne in mind when planning an efficient synthesis of a substituted benzene.

(1) The introduction of an *activating group* onto a benzene ring makes the product more reactive than the starting material. It is therefore difficult to stop after the first substitution. This is observed in Friedel–Crafts alkylations.

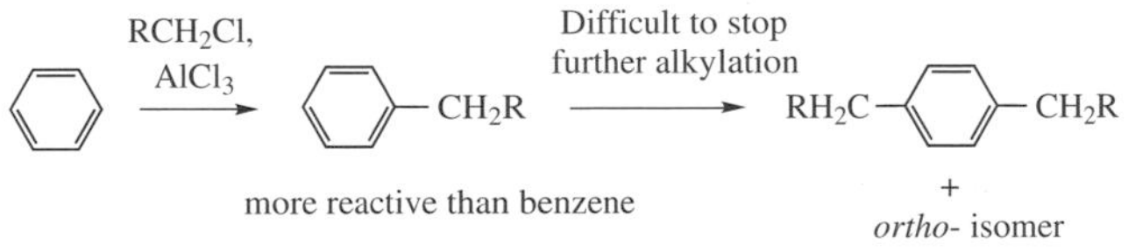

(2) The introduction of a *deactivating group* onto a benzene ring makes the product less reactive than the starting material. Therefore, multiple substitutions do not occur, for example, in Friedel–Crafts acylations (to give ketones). The ketone (ArCOR) could then be reduced to give the mono-alkylated product (ArCH$_2$R) in

generally higher yield than that derived from a Friedel–Crafts alkylation.

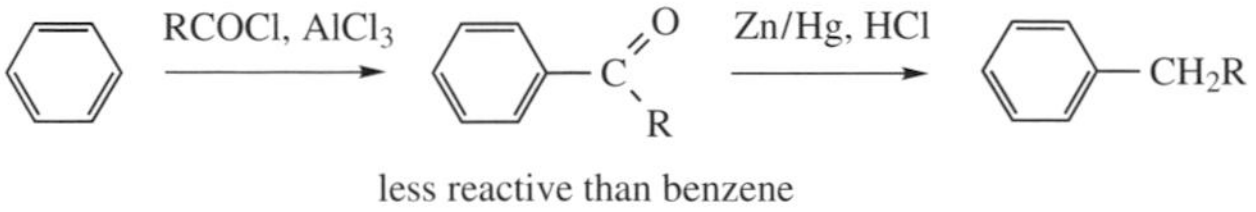

less reactive than benzene

(3) For an *ortho-/para*-substituted benzene, the first group to be introduced on the ring should be *ortho-/para-* directing. For a *meta*-substituted benzene ring, the first group to be introduced on the ring should be *meta-* directing.

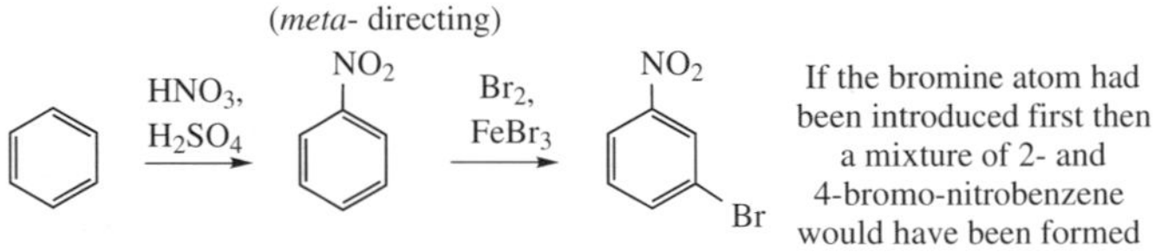

(*meta-* directing)

If the bromine atom had been introduced first then a mixture of 2- and 4-bromo-nitrobenzene would have been formed

(4) For aniline, the orientation of substitution depends on the pH of the reaction. At low pH, aniline is protonated, and the protonated amine is *meta-* directing.

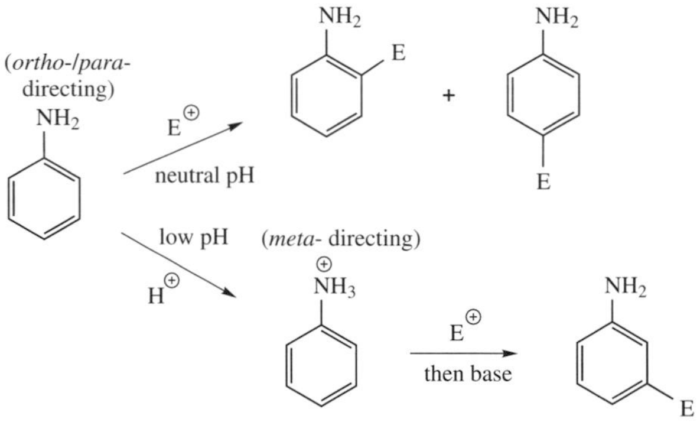

(*ortho-/para-* directing)

neutral pH

low pH   (*meta-* directing)

then base

(5) Making the activating group larger in size can increase the proportion of *para-* over *ortho-* substitution (as steric hindrance minimises *ortho-* attack). For example, the conversion of an amine to a bulkier amide leads to *para-* substitution, and the amide can be described as a 'blocking group'. The amide group is also not as strong an activating group as the amine (see Section 7.3.1), and hence multiple substitution is less of a problem.

(*ortho-/para-*
directing)

NH₂ → CH₃COCl / Base → NHCOR → E⁺ → NHCOR (para-) → NaOH, H₂O / Hydrolysis of amide back to amine → NH₂

Reaction with an electrophile gives a mixture of *ortho-/para-* isomers

amide substituent is larger than the amine

para- isomer

(6) Removable aromatic substituents, typically SO₃H or NH₂, can be used to direct or block a particular substitution. In this way, aromatic products with unusual substitution patterns can be prepared.

(*ortho-/para-*
directing)

(1) H₂SO₄, HNO₃ (2) Sn, HCl → NH₂ → Br₂, FeBr₃ → NH₂ (with Br groups) → (1) NaNO₂, HCl (2) H₃PO₂ → (Br groups meta)

Amine group is removed leaving Br groups *meta-* to one another

(7) Deactivating groups can be converted to activating groups (and vice versa) by functional group interconversion.

(*ortho-/para-*
directing) CH₃ → KMnO₄ → (*meta-*
directing) CO₂H   (*meta-*
directing) NO₂ → Sn, HCl → (*ortho-/para-*
directing) NH₂

(8) For the formation of trisubstituted benzenes, the directing effects of the two groups on the benzene ring must be compared. When these groups direct to different positions, the more powerful activating group usually has the dominant influence. Substitution in between two *meta*-disubstituted groups is rare because of steric hindrance.

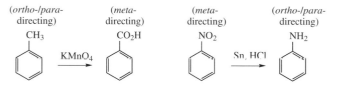

(*ortho-/para-*
directing) CH₃ ... E⁺ / CO₂H (*meta-*
directing)

(*ortho-/para-*
directing) OH ... E⁺ / CH₃ (*ortho-/para-*
directing) — OH is a more powerful activating group

Attack at this site is hindered — (*ortho-/para-*
directing) CH₃ ... E⁺ / Cl (*ortho-/para-*
directing) E⁺

7.9   Electrophilic substitution of naphthalene

As naphthalene is aromatic (ten π-electrons), it also undergoes electrophilic substitution reactions. Attack occurs selectively at C-1 rather than C-2, because the intermediate carbocation is more stable (i.e. two resonance structures can be drawn with an intact benzene ring. For attack at C-2, only one can be drawn with an intact benzene ring).

For sulfonation (using H_2SO_4), the position of attack depends on the reaction temperature. At 80 °C, attack at C-1 occurs (as expected), and 1-sulfonic acid is formed under kinetic control (see Section 4.9.3). However, at higher temperatures (e.g. 160 °C), attack at C-2 predominates, and the 2-sulfonic acid is formed under thermodynamic control. The C-1 isomer is thermodynamically less stable than the C-2 isomer because of an unfavourable steric interaction with the hydrogen atom at C-8 (this is called a *peri interaction*).

7.10   Electrophilic substitution of pyridine

Pyridine, like benzene, has six π-electrons. The electron-withdrawing nitrogen atom deactivates the ring, and electrophilic substitution is slower than that for benzene. Substitution occurs principally at the 3-position of the ring, as attack at the 2-/4-position produces less stable cation intermediates (i.e. with one resonance structure having a positive charge on the divalent nitrogen).

Attack at C-2
Unfavourable

Attack at C-4
Unfavourable

Attack at C-3
Preferred

Nitrogen lone pair is not
part of the aromatic sextet

7.11   Electrophilic substitution of pyrrole, furan and thiophene

In pyrrole, the lone pair of electrons on nitrogen is part of the aromatic (six π-electron) ring system. The incorporation of a lone pair of electrons activates the ring, and electrophilic substitution is faster than that for benzene. Substitution occurs principally at the 2-position, as attack at the 3-position produces a less stable cation intermediate (i.e. with two, rather than three, resonance structures).

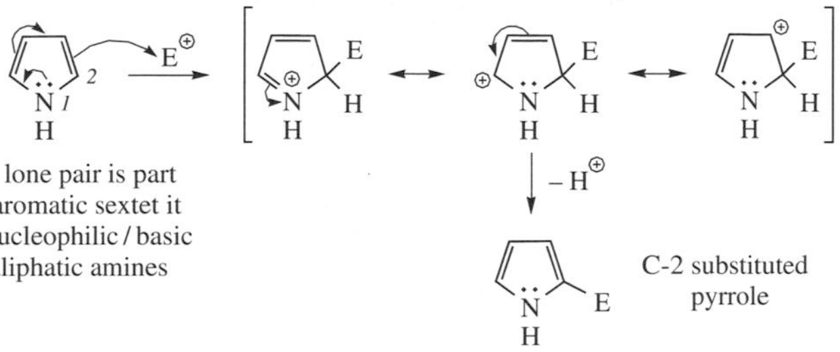

As the lone pair is part of the aromatic sextet it is less nucleophilic / basic than aliphatic amines

C-2 substituted pyrrole

Furan and thiophene also undergo electrophilic substitution reactions, although not so readily as pyrrole. The typical order of reactivity to electrophiles is: pyrrole > furan > thiophene > benzene. Pyrrole is most reactive because the nitrogen atom is a more powerful electron donor than the oxygen atom in furan (i.e. nitrogen is less electronegative than oxygen). Thiophene is less reactive than either pyrrole or furan because the lone pair of electrons on sulfur is in a 3p- (rather than a 2p) orbital. The 3p-orbitals of sulfur overlap less efficiently with the 2p-orbitals on carbon than the (similar size) 2p-orbitals of nitrogen or oxygen. As for pyrrole, the 2-position of both furan and thiophene is more reactive to electrophiles than the 3-position.

Examples

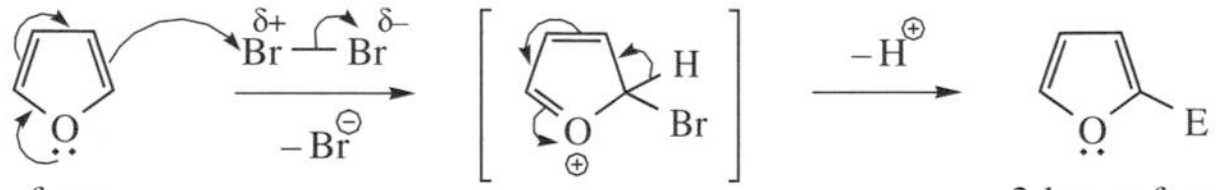

furan

Furan is more reactive to electrophiles than benzene and hence FeBr$_3$ is not required

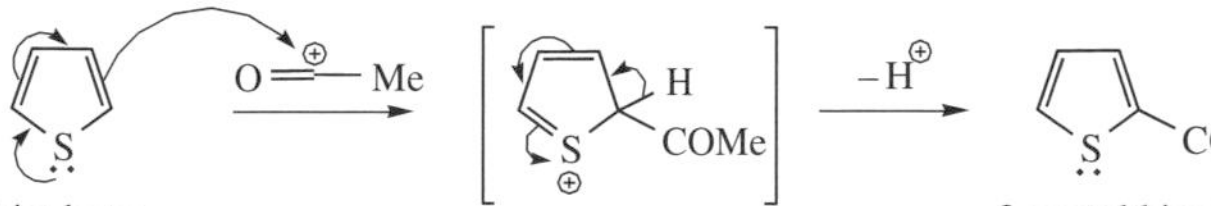

2-bromofuran

thiophene

The acylium ion can be prepared using MeCOCl and SnCl$_4$

2-acetylthiophene

Problems

(1)   Give the mechanism of the electrophilic substitution reaction of benzene with an electrophile E$^+$.

(2)   Write a mechanism for the formation of (1,1-dimethylpropyl)-benzene from 1-chloro-2,2-dimethylpropane, benzene and a catalytic amount of AlCl$_3$.

(3) (a) On the basis of Hückel's rule, label the following molecules
A–D as aromatic or anti-aromatic?

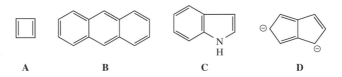

A   **B**   **C**   **D**

(b) Describe one experimental approach for testing the aromatic character of **C**.

(4) Suggest syntheses of the following compounds starting from benzene.
(a) Triphenylmethane
(b) (1-Bromopropyl)benzene
(c) 4-Bromobenzoic acid
(d) 2,4-Dinitrophenylhydrazine

(5) Explain the following:
(a) Reaction of phenylamine (aniline) with bromine yields 2,4,6-tribromophenylamine, whereas nitration (with a mixture of concentrated nitric and hydrochloric acids) gives mainly 3-nitrophenylamine.
(b) Bromination of benzene with Br_2 requires $FeBr_3$, whereas reaction of Br_2 with cyclohexene or phenol does not.
(c) Although 4-bromophenylamine can be formed in one step from phenylamine, the yield is low and hence a three-step synthesis involving the formation of an intermediate amide is preferred.

(6) How would you prepare the following compounds from phenylamine (aniline) using arenediazonium salts as intermediates?
(a) Iodobenzene
(b) 4-Bromochlorobenzene
(c) 1,3,5-Tribromobenzene

8. CARBONYL COMPOUNDS: ALDEHYDES AND KETONES

Key point. The carbonyl (C=O) bond is polarised, and the oxygen atom is slightly negative, while the carbon atom is slightly positive. This leads to the addition of nucleophiles to the carbon atom of aldehydes or ketones in *nucleophilic addition reactions*. A variety of charged or neutral nucleophiles can add to the carbonyl group, although addition of neutral nucleophiles usually requires the presence of an acid catalyst. Aldehydes and ketones bearing α-hydrogen atoms can undergo tautomerism to form *enols* or, on reaction with a base, undergo deprotonation to form *enolates*. Enols and particularly enolates can act as nucleophiles, and reaction with an electrophile leads to an *α-substitution reaction*. When an enolate reacts with an aldehyde or ketone, this can produce an *enone* in a *carbonyl–carbonyl condensation reaction*.

8.1 Structure

All carbonyl compounds contain an acyl fragment (RCO) bonded to another residue. The carbonyl carbon atom is sp^2 hybridised (three σ-bonds and one π-bond), and as a consequence, the carbonyl group is planar and has bond angles of around 120°. The C=O bond is short (1.22 Å) and also rather strong (ca. 690 kJ mol^{-1}).

π-bond

two lone pairs

R^1 $\delta+$ $\delta-$
C=O
R

Ketone: R = R^1 = alkyl or aryl
Aldehyde: R = alkyl or aryl, R^1 = H

As oxygen is more electronegative than carbon, the electrons in the C=O bond are drawn towards the oxygen. This means that carbonyl compounds are polar and have substantial dipole moments.

Carbonyl compounds show characteristic peaks in the infrared and ^{13}C NMR spectra (see Sections 10.4 and 10.5.2, respectively). A characteristic signal is also observed for the aldehyde proton in the ^{1}H NMR spectrum (see Section 10.5.1).

8.2 Reactivity

Polarisation of the C=O bond means that the carbon atom is electrophilic (δ+) and the oxygen atom is nucleophilic (δ−). Therefore, nucleophiles attack the carbon atom and electrophiles attack the oxygen atom.

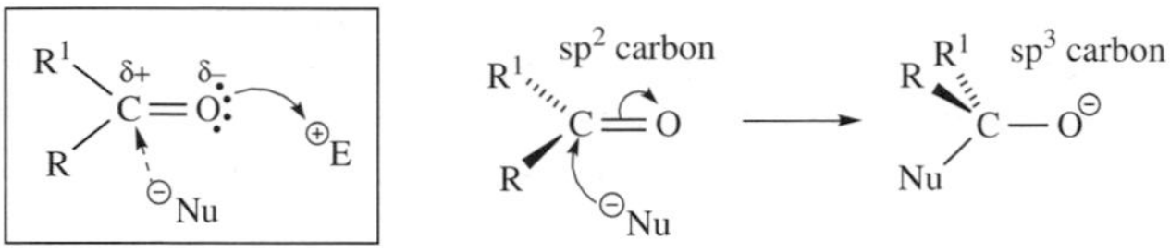

Attack by a nucleophile breaks the π-bond, and the electrons reside on the oxygen atom. This is energetically favourable, as a strong σ-bond is formed at the expense of a weaker π-bond. Examination of crystal structures has shown that the nucleophile approaches the carbonyl carbon at an angle of around 107°. This is known as the *Bürgi–Dunitz angle* (see Section 4.10).

There are three general mechanisms by which aldehydes and ketones react. These are (i) *nucleophilic addition reactions*, (ii) *α-substitution reactions* and (iii) *carbonyl–carbonyl condensation reactions*.

(1) *Nucleophilic addition reactions.* Both charged and uncharged nucleophiles can attack the carbonyl carbon atom to form addition products. This is the most common reaction for aldehydes and ketones.

With charged nucleophiles

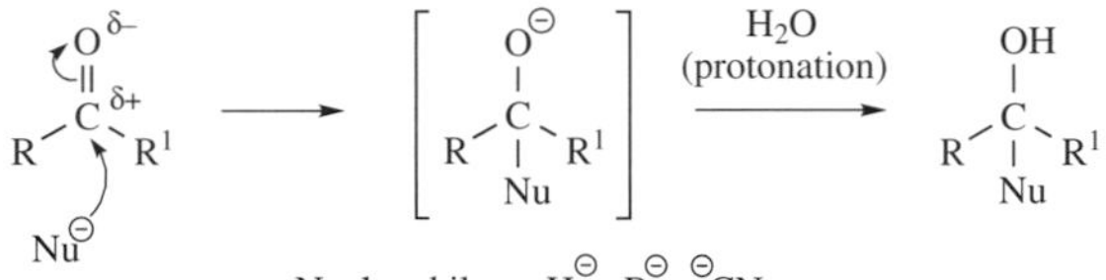

Nucleophiles = $H^{\ominus}$, $R^{\ominus}$, $^{\ominus}CN$

With uncharged nucleophiles

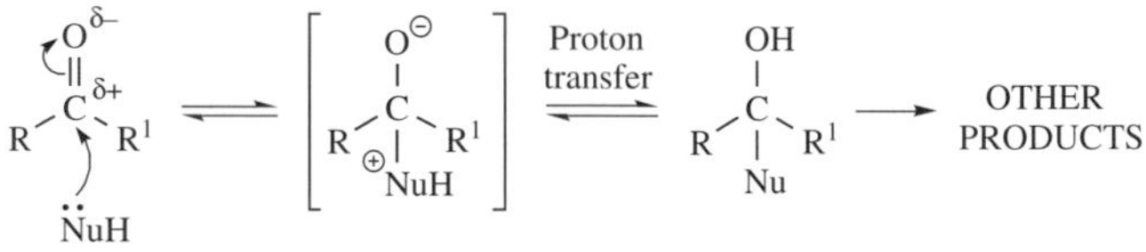

Nucleophiles = H_2O, ROH, RSH, NH_3, RNH_2, R_2NH
This process can be acid catalysed

(2) *α-Substitution reactions.* This involves reaction at the position next to the carbonyl group, which is known as the α-position. Deprotonation produces an enolate, which can act as a nucleophile.

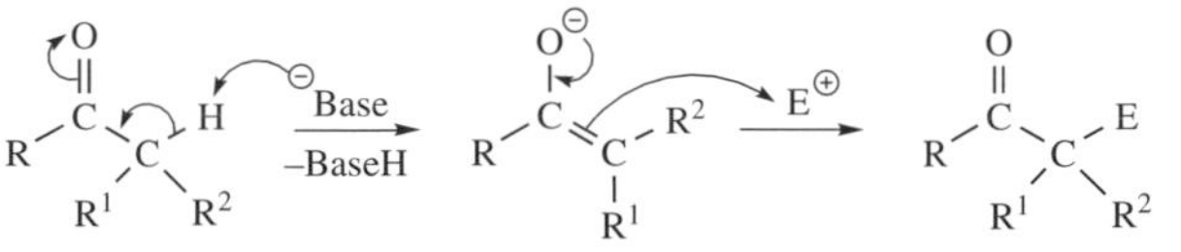

The carbonyl group renders the hydrogen(s) on the α-carbon acidic

An enolate ion

(3) *Carbonyl–carbonyl condensation reactions.* These reactions involve both a nucleophilic addition step and an α-substitution step. They can occur when two carbonyl compounds react with one another. For example, reaction of two molecules of ethanal.

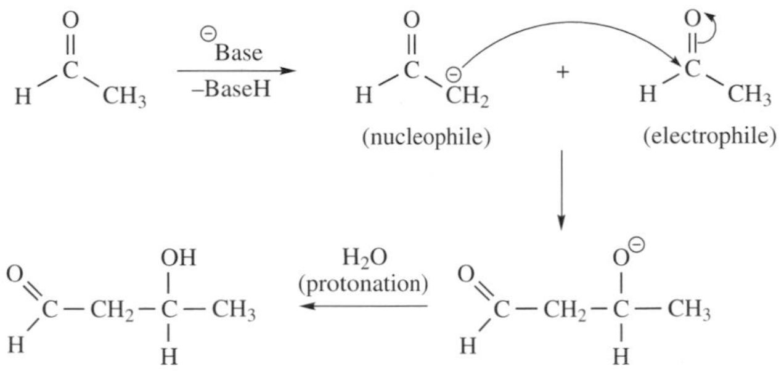

8.3 Nucleophilic addition reactions

8.3.1 Relative reactivity of aldehydes and ketones

Aldehydes are generally more reactive than ketones for both steric and electronic reasons.

(1) *On steric grounds.* The nucleophile can attack the aldehyde carbonyl carbon atom more readily as this has only one (rather than two) alkyl group(s) bonded to it. The transition state resulting from addition to the aldehyde is less crowded and lower in energy.

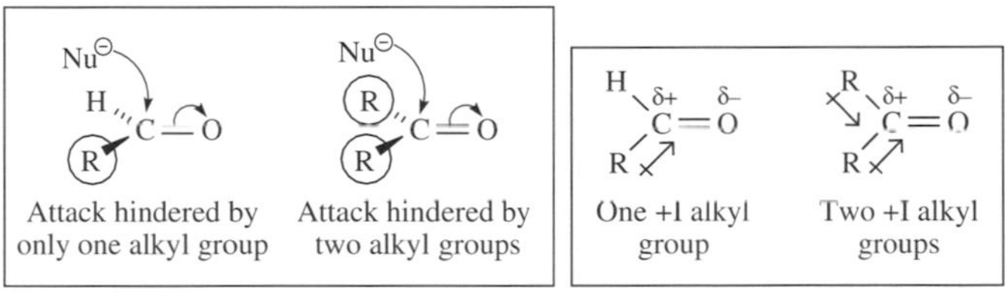

| Attack hindered by only one alkyl group | Attack hindered by two alkyl groups | One +I alkyl group | Two +I alkyl groups |

(2) *On electronic grounds.* Aldehydes have a more electrophilic carbonyl carbon atom because there is only one (rather than two) alkyl group(s) that donates electron density towards it.

8.3.2 Types of nucleophiles

The attacking nucleophile can be negatively charged (Nu^-) or neutral (Nu). With neutral nucleophiles, acid catalysis is common.

$Nu^- = H^-$ (hydride), R^- (carbanion), NC^- (cyanide), HO^- (hydroxide), RO^- (alkoxide).

$Nu = H_2O$ (water), ROH (alcohol), NH_3 (ammonia), RNH_2 or R_2NH (primary or secondary amine).

8.3.3   Nucleophilic addition of hydride: reduction

Reduction of aldehydes and ketones with hydride leads to the formation of alcohols. The hydride ion can be derived from a number of reagents.

8.3.3.1   Complex metal hydrides

Lithium aluminium hydride ($LiAlH_4$) or sodium borohydride ($NaBH_4$) can act as hydride donors. A *simplified* view of the mechanism involves the formation of an intermediate alkoxide, which on protonation yields a primary or secondary alcohol.

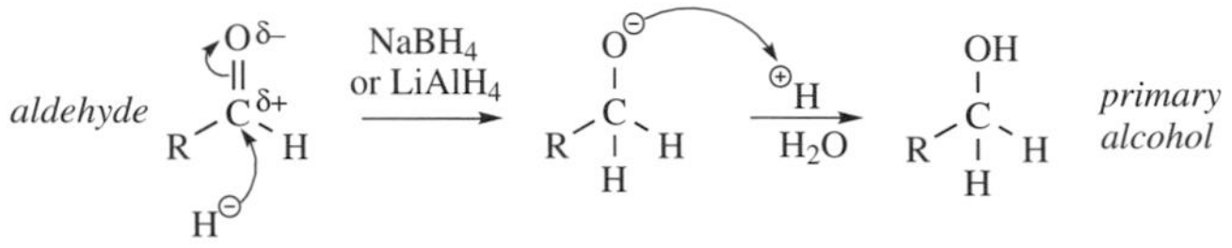

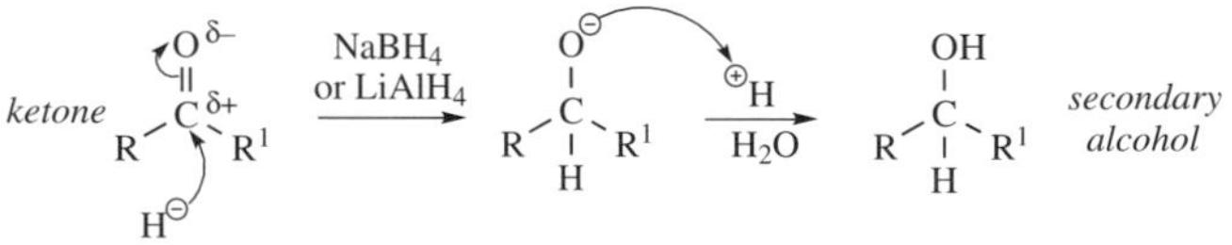

The metal–hydrogen bonds in $LiAlH_4$ are more polar than those in $NaBH_4$. As a consequence, $LiAlH_4$ is a stronger reducing agent than $NaBH_4$.

8.3.3.2   The Meerwein–Ponndorf–Verley reaction

The hydride anion is derived from 2-propanol, which is first deprotonated to form an alkoxide.

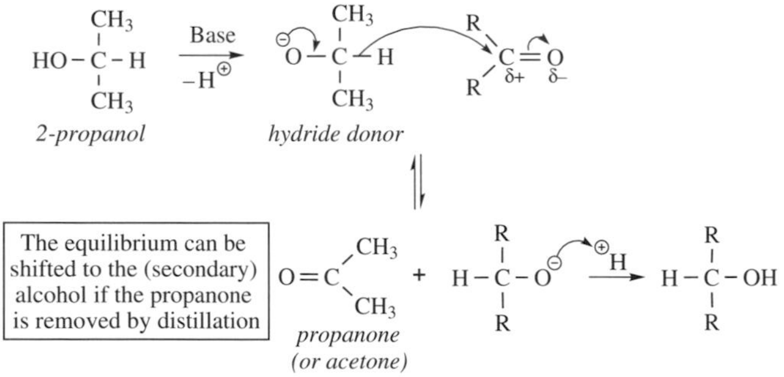

Lewis acids, such as $Al(OR)_3$, are employed so as to facilitate hydride transfer by forming a complex with the alkoxide and carbonyl compound.

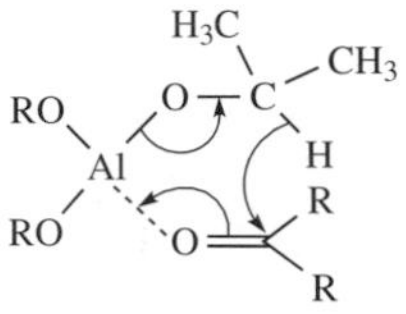

The reverse reaction, involving the oxidation of the secondary alcohol to form a ketone, is known as the *Oppenauer oxidation*. In this case, the equilibrium is pushed to the ketone by using excess propanone.

8.3.3.3 The Cannizzaro reaction

The hydride anion is derived from aldehydes such as methanal (HCHO), which do not contain an α-hydrogen atom (i.e. a hydrogen atom on the carbon atom next to the carbonyl).

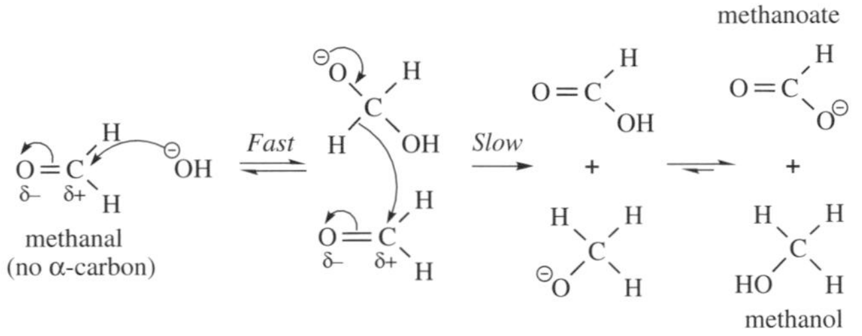

This is a *disproportionation reaction*, as one molecule of methanal is oxidised to methanoate (which on protonation gives methanoic acid) and the other molecule is reduced to methanol.

8.3.3.4 Nicotinamide adenine dinucleotide (NADH)

Similar hydride reductions occur in nature using NADH in the presence of enzymes.

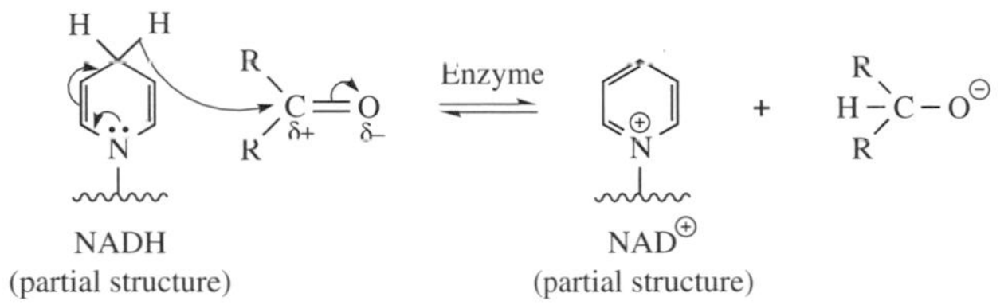

8.3.3.5 The reverse of hydride addition: oxidation of alcohols

The conversion of *secondary* alcohols (R_2CHOH) to ketones can be achieved by $KMnO_4$, $Na_2Cr_2O_7$ or CrO_3 under acidic conditions. The use of CrO_3 in acid is known as the *Jones oxidation*.

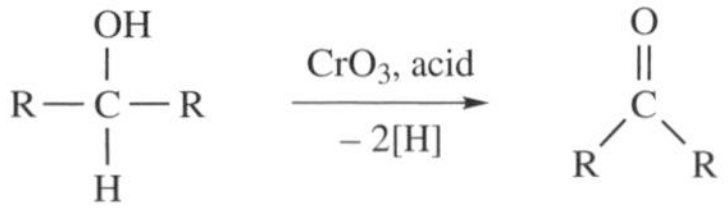

The mechanism involves the formation of a chromate ester, and at the end of the reaction, the oxidation state of the chromium changes from +6 (orange colour) to +3 (green colour), i.e. the chromium is reduced.

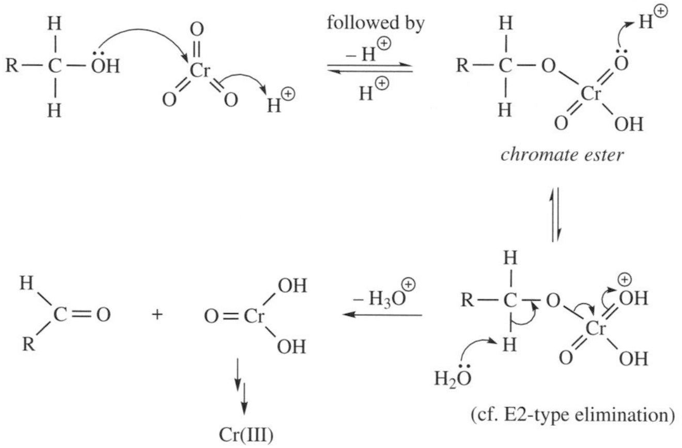

chromate ester

(cf. E2-type elimination)

Cr(III)

For *primary* alcohols (RCH_2OH), it is often difficult to stop at the aldehyde stage, and further oxidation to give the carboxylic acid can occur. The aldehyde can be isolated by distillation (i.e. removed from the reaction mixture as soon as it is formed).

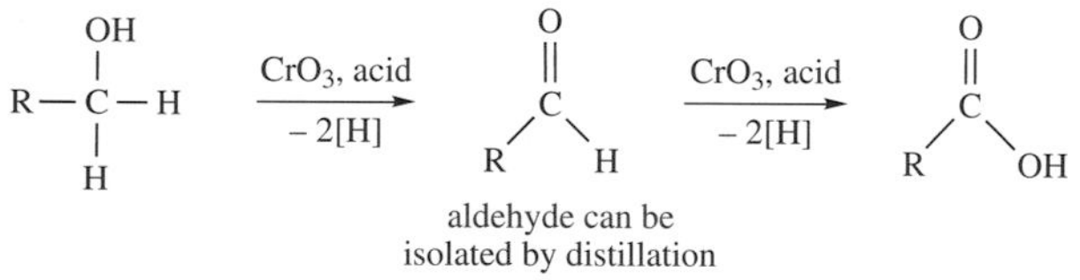

aldehyde can be isolated by distillation

Oxidising agents milder than CrO_3/H^+, including pyridinium chlorochromate ($C_5H_5NH^+$ $ClCrO_3^-$), abbreviated as PCC, can also be used. PCC is an excellent reagent for oxidising primary alcohols to aldehydes in anhydrous dichloromethane.

Tertiary alcohols (R_3C-OH) cannot be oxidised by CrO_3/H^+ (or related reagents) because there are no hydrogen atoms on the carbon atom bearing the OH group. As a result, oxidation must take place by breaking the carbon–carbon bonds (e.g. in combustion, to give CO_2 and H_2O).

8.3.4 Nucleophilic addition of carbon nucleophiles: formation of C–C bonds

8.3.4.1 Reaction with cyanide

Addition of cyanide leads to the reversible formation of a cyanohydrin. Only a catalytic amount of the cyanide anion in the presence of hydrogen cyanide is required (as the cyanide anion is regenerated).

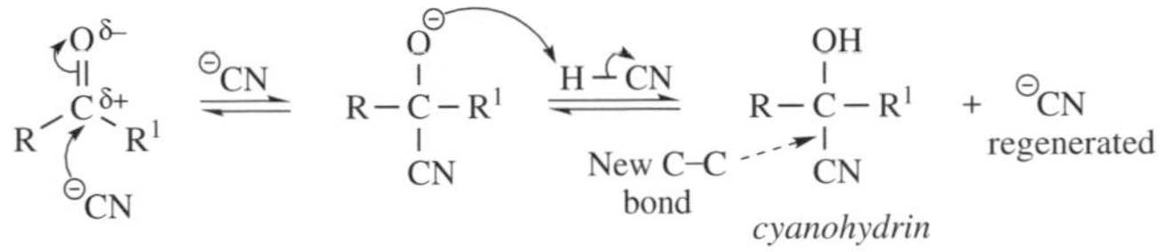

cyanohydrin

The position of the equilibrium depends on the structure of the carbonyl compound. Nucleophilic addition is favoured by small alkyl groups attached to the carbonyl group and also by electron-withdrawing groups (e.g. CCl_3), which increase the δ+ character of the carbonyl carbon atom (see Section 8.3.1).

Cyanohydrins are useful because they can be converted into other functional groups, e.g. hydroxy acids (see Section 9.9 for the reaction mechanism).

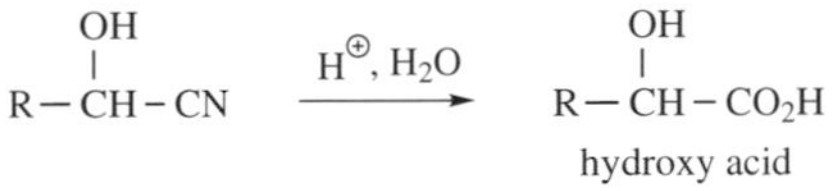

hydroxy acid

The reaction of cyanide with *aromatic* aldehydes leads to the *benzoin condensation reaction*. The product from this reaction is called benzoin.

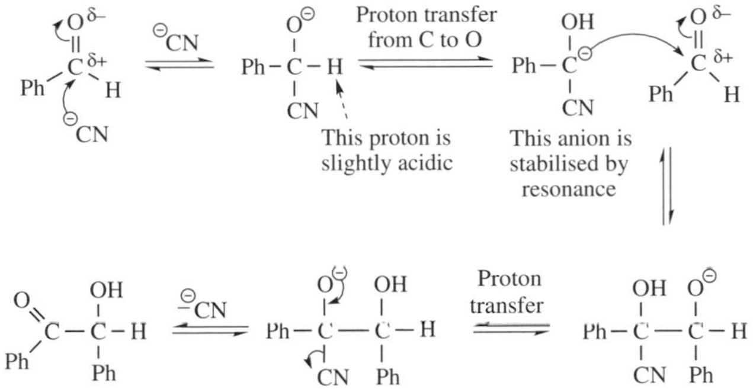

8.3.4.2 Reaction with organometallics

Organometallic compounds (R—Metal) are a source of nucleophilic alkyl or aryl groups. This is because the metal is more electropositive than the carbon atom to which it is bonded.

$$\overset{\delta-}{R}\text{——}\overset{\delta+}{\text{Metal}}$$

Organometallic reagents include:

(1) organolithium reagents, R—Li;
(2) Grignard reagents, R—MgX (where X = Cl, Br or I);

(3) the Reformatskii reagent, $BrZn{-}CH_2CO_2Et$;
(4) lithium alkynides, $RC{\equiv}C{-}Li$.

The nucleophilic alkyl or aryl group adds to the carbonyl to form an alcohol. These reactions are very useful in the synthesis of complex organic molecules because new carbon–carbon bonds are formed.

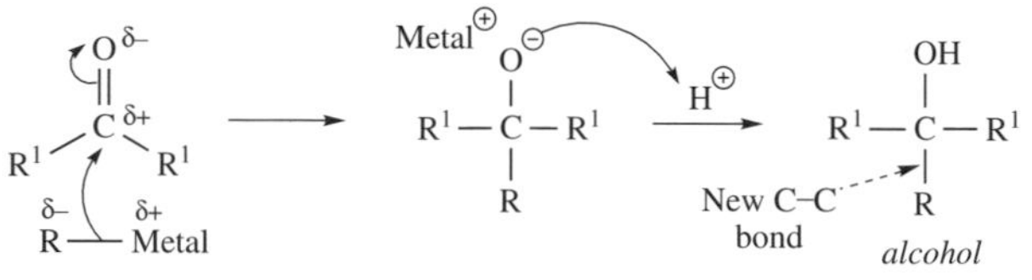

Examples

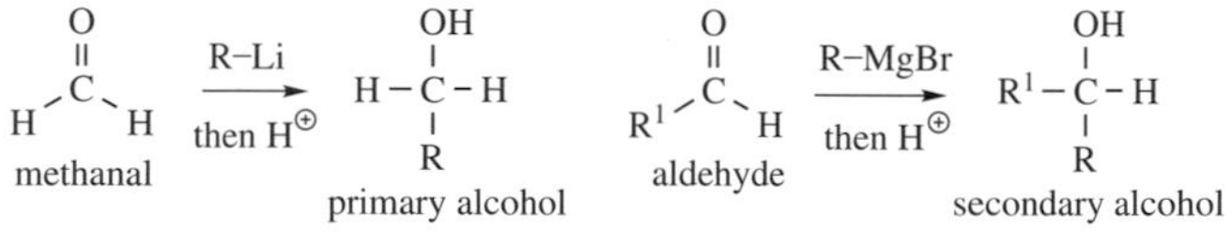

Organolithium and Grignard reagents are prepared from alkyl (or aryl) halides. The reactions are carried out under anhydrous conditions because reaction with water leads to the formation of alkanes.

$$R{-}X \quad + \quad 2Li \xrightarrow{\text{dry pentane}} R{-}Li \quad + \quad LiX$$

$$R{-}X \; + \; Mg \xrightarrow{\text{dry diethyl ether}} R{-}MgX \xrightarrow{H{-}OH} \underset{\textit{alkane}}{R{-}H} \; + \; HOMgX$$

8.3.4.3 Reaction with phosphorus ylides: the Wittig reaction

Aldehydes and ketones react with ylides or phosphoranes ($R_2C{=}PPh_3$) to form alkenes. The first step involves nucleophilic attack by the carbon atom of the phosphorane.

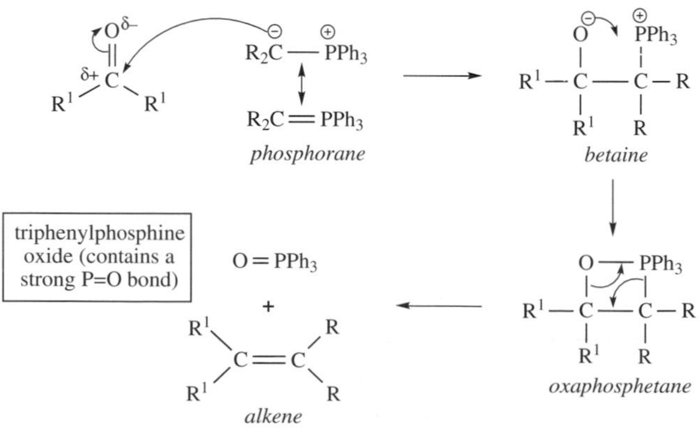

The driving force for the reaction is the formation of the very strong P=O bond in triphenylphosphine oxide.

Phosphorus ylides are prepared from a nucleophilic substitution reaction between alkyl halides and triphenylphosphine (PPh₃). The resulting alkyltriphenylphosphonium salt is then deprotonated by reaction with a strong base (e.g. BuLi) to form the ylide.

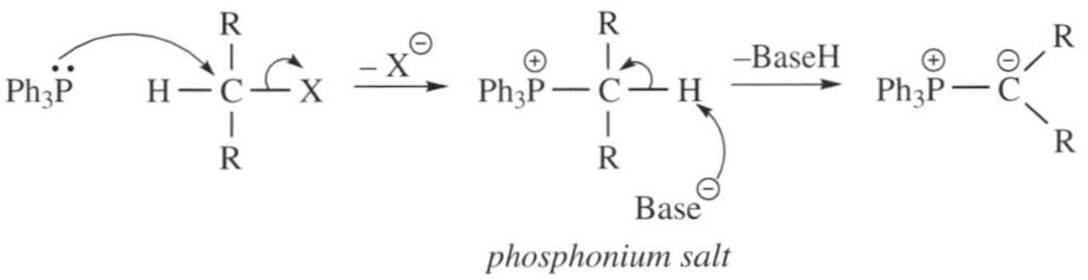

phosphonium salt

8.3.5 Nucleophilic addition of oxygen nucleophiles: formation of hydrates and acetals

8.3.5.1 Addition of water: hydration

Aldehydes and ketones undergo a reversible reaction with water to yield 1,1-diols (otherwise known as geminal (gem) diols or hydrates).

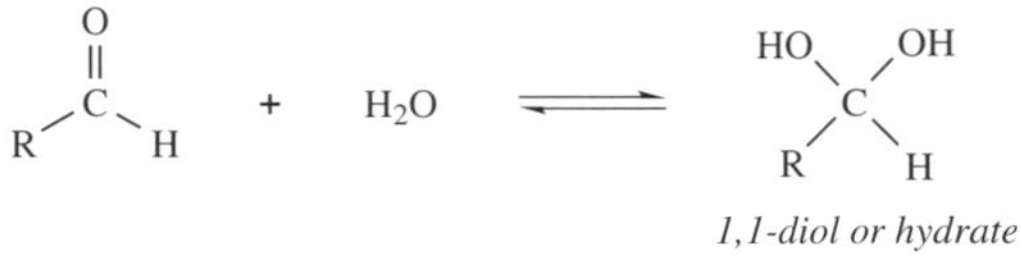

1,1-diol or hydrate

The addition of water is slow but can be catalysed by bases or acids.

Base catalysis

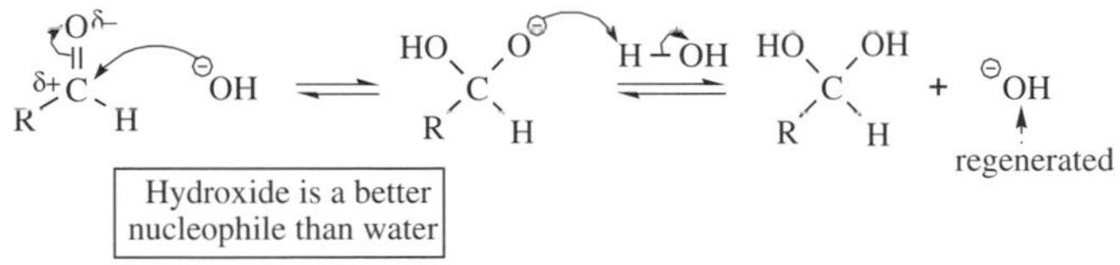

Hydroxide is a better nucleophile than water

Acid catalysis

The carbonyl acts as a base

The protonated carbonyl is a better electrophile

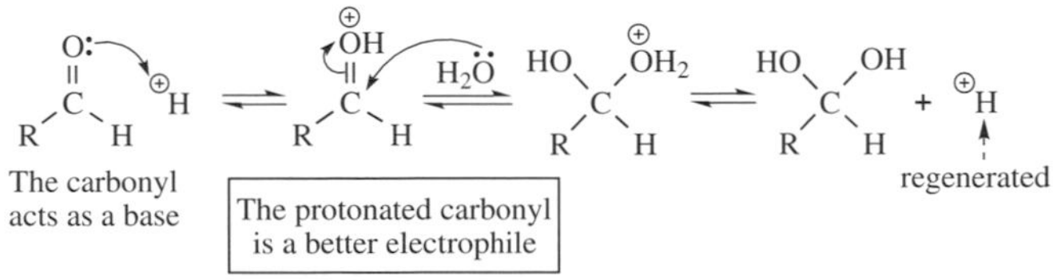

Electron-donating and bulky substituents attached to the carbonyl group decrease the percentage of 1,1-diol present at equilibrium, whereas electron-withdrawing and small substituents increase it. Therefore, only 0.2% of propanone ($O=CMe_2$) is hydrated at equilibrium, while 99.9% of methanal ($O=CH_2$) is hydrated.

8.3.5.2   Addition of alcohols: hemiacetal and acetal formation

Alcohols are relatively weak nucleophiles (like water) but add rapidly to aldehydes and ketones on acid catalysis. The initial product is a hydroxyether or *hemiacetal*.

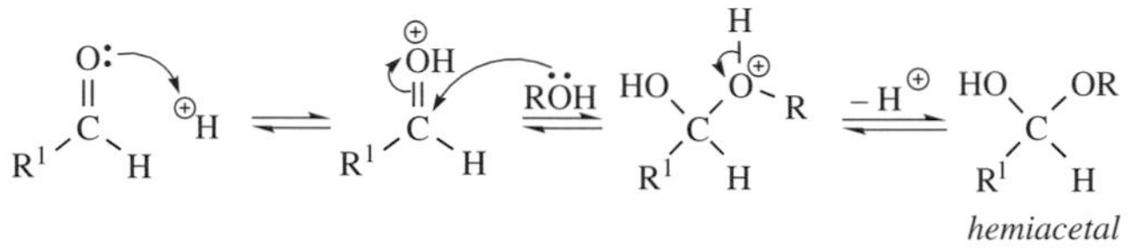

Hemiacetal formation is fundamental to the chemistry of carbohydrates (see Section 11.1). Glucose, for example, contains an aldehyde and several alcohol groups. The reaction of the aldehyde with one of the alcohols leads to the formation of a cyclic hemiacetal (even without acid catalysis) in an intramolecular reaction.

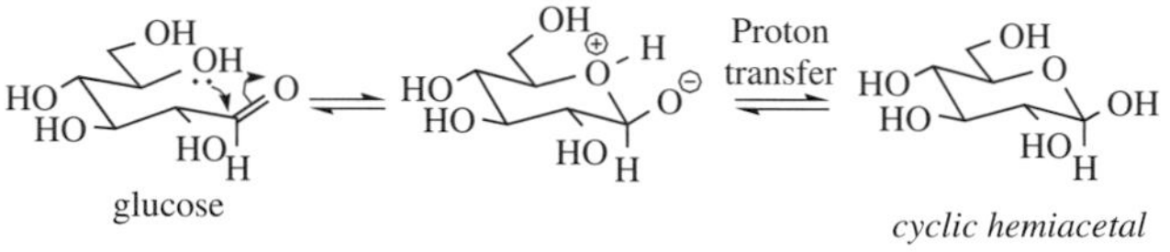

Hemiacetals can undergo further reaction, with a second equivalent of alcohol, to yield an *acetal*.

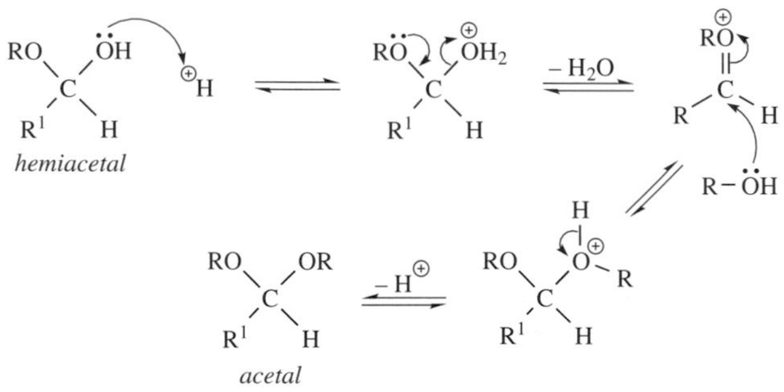

The whole sequence of acetal formation is *reversible*.

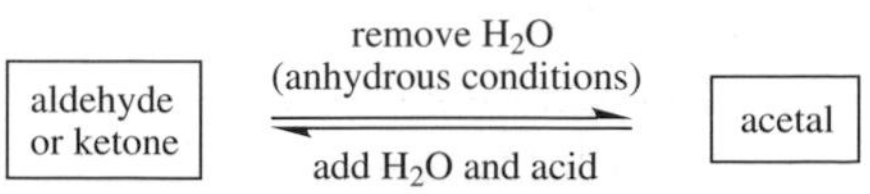

Acetals are extremely useful compounds in synthesis because they can act as *protecting groups* for aldehydes and ketones.

Example

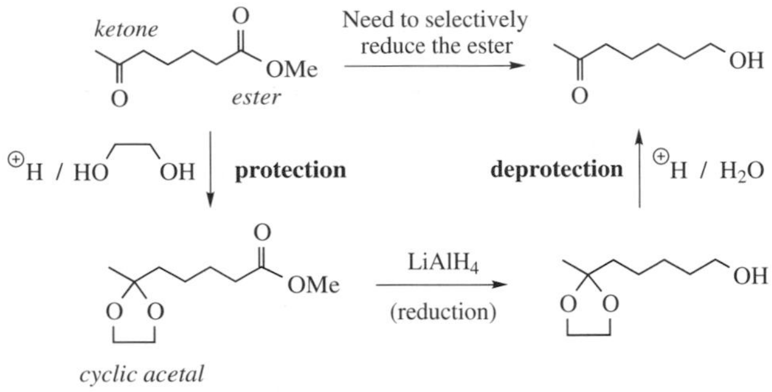

cyclic acetal

8.3.6 Nucleophilic addition of sulfur nucleophiles: formation of thioacetals

Thiols (RSH) add reversibly to aldehydes and ketones in the presence of acid to yield *thioacetals*. The reaction mechanism is analogous to the formation of acetals.

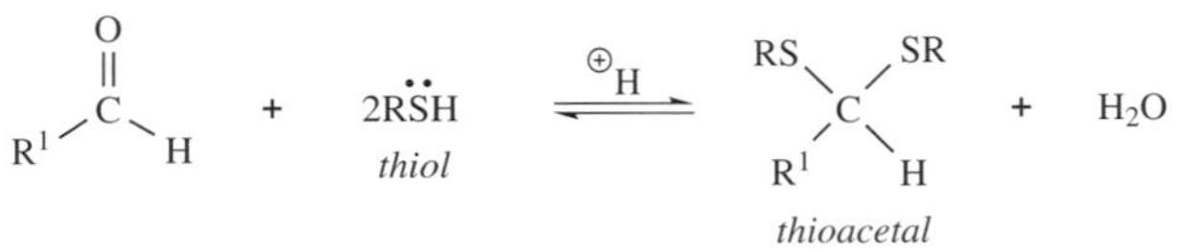

Thioacetals are useful because they undergo desulfurisation when treated with Raney nickel in the *Mozingo reduction*. This is an excellent method for reducing aldehydes or ketones to alkanes.

Example

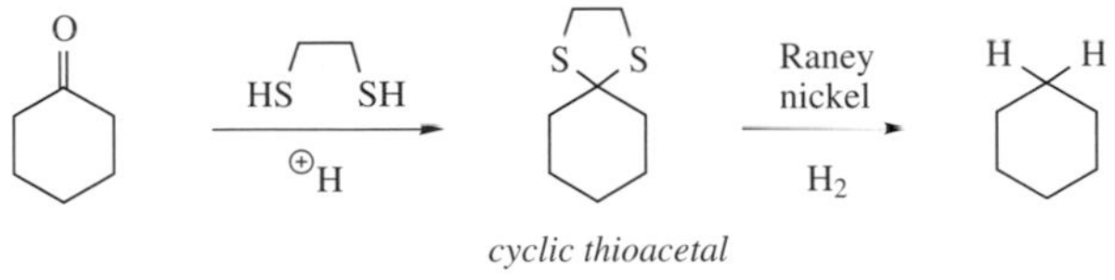

cyclic thioacetal

8.3.7 Nucleophilic addition of amine nucleophiles: formation of imines and enamines

Aldehydes and ketones react with primary amines (RNH_2) to yield *imines* and react with secondary amines (R_2NH) to form *enamines*.

8.3.7.1 Formation of imines

This reversible, acid-catalysed process involves the addition of a nucleophile, followed by elimination of water; this is known as an *addition–elimination* or *condensation* reaction. (A condensation reaction involves the combination of two molecules to form one larger molecule, with the elimination of a small molecule, often water.)

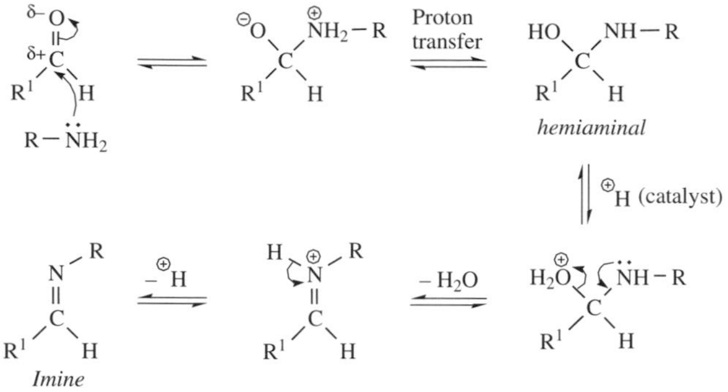

The reaction is pH dependent.

- At *low* pH (strongly acidic conditions), the amine is protonated (RNH_3^+) and hence cannot act as a nucleophile.
- At *high* pH (strongly alkaline conditions), there is not enough acid to protonate the OH group of the hemiaminal to make this a better leaving group.

The best compromise is around pH 4.5.

There are a number of related reactions, which employ similar nucleophiles.

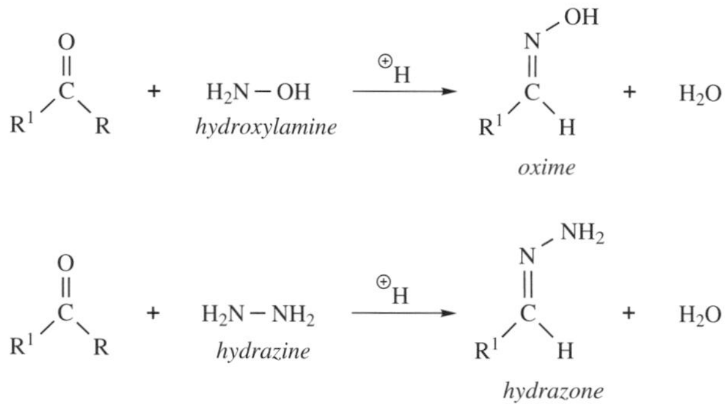

8.3.7.2   Reactions of imines, oximes and hydrazones

The most important reaction of imines is their reduction to form amines. The conversion of an aldehyde or ketone to an amine, *via* an imine, is known as *reductive amination*. This allows, for example, the selective formation of a secondary amine from a primary amine.

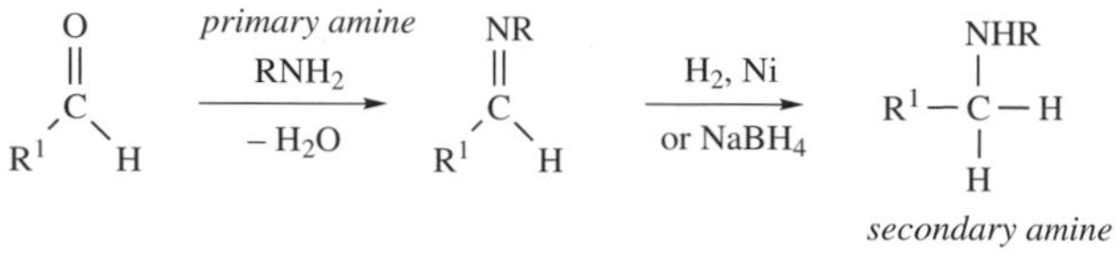

Imines can also be attacked by nucleophiles, such as cyanide, and this is used in the *Strecker amino acid synthesis*.

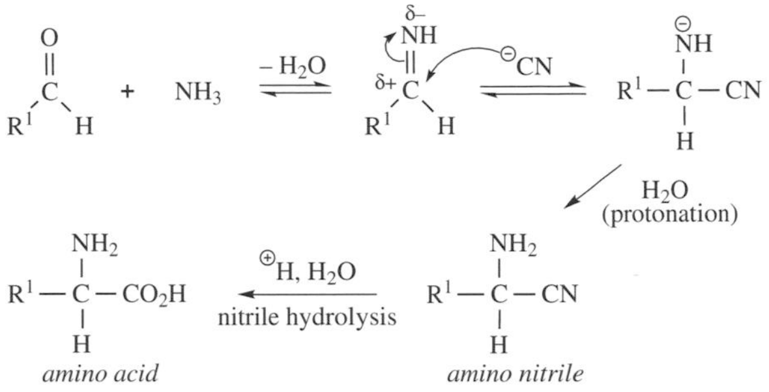

An important reaction of oximes is their conversion to amides in the *Beckmann rearrangement reaction*.

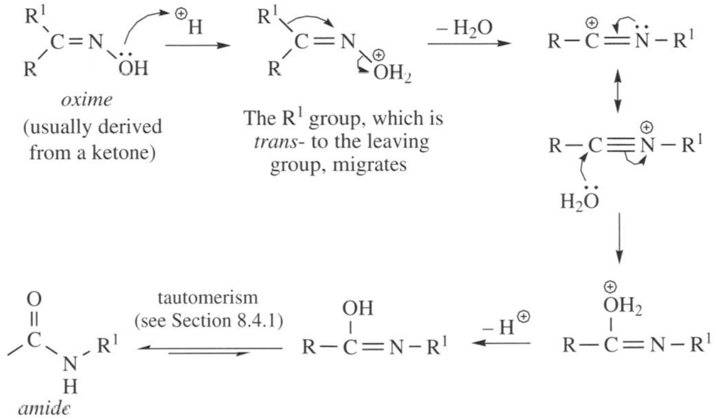

An important reaction of hydrazones is their conversion to alkanes on heating with hydroxide. This is known as the *Wolff–Kischner reaction*.

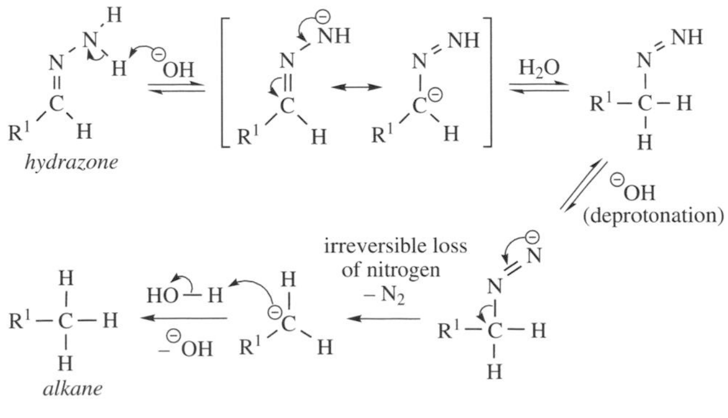

An aldehyde or ketone can therefore be reduced to an alkane *via* a hydrazone. This transformation can also be achieved using the *Mozingo reduction* (see Section 8.3.6) or the *Clemmensen reduction* (shown below; see Section 7.6).

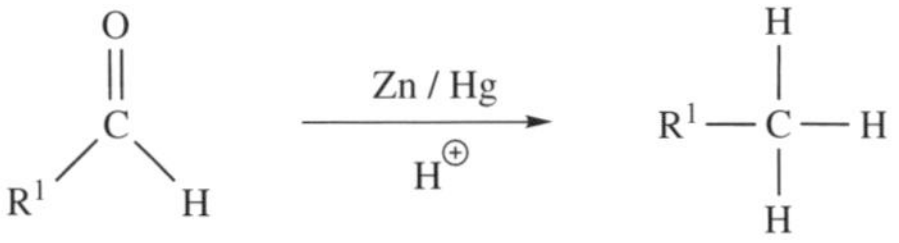
Zn / Hg
H⊕

8.3.7.3 Formation of enamines

Enamines are prepared by the reaction of a ketone or aldehyde with a *secondary* amine.

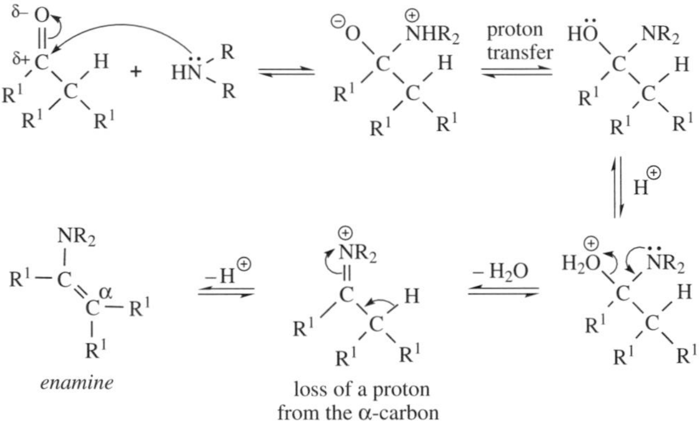
proton transfer
− H₂O
−H⊕
H⊕
enamine
loss of a proton from the α-carbon

8.4 α-Substitution reactions

These reactions take place at the position next to the carbonyl group (i.e. the α-position) and involve substitution of an α-H atom by another group. The reactions take place *via enol* or *enolate* intermediates.

8.4.1 Keto–enol tautomerism

Carbonyl (or keto) compounds are interconvertible with their corresponding enols. This rapid interconversion of structural isomers under ordinary conditions is known as *tautomerism*. Keto–enol tautomerism is catalysed by acids or bases.

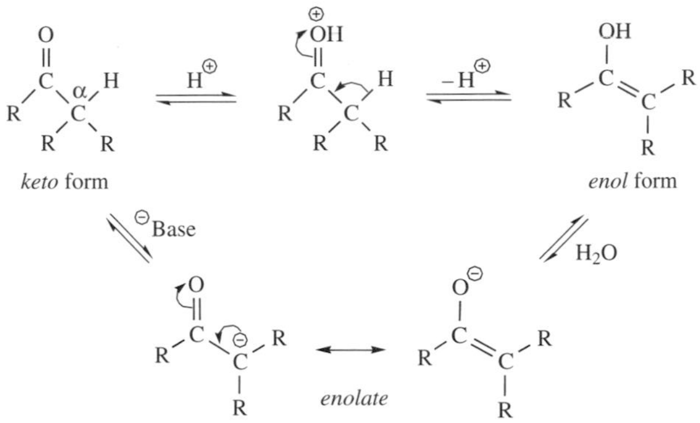

keto form   *enol* form

enolate

This is an example of *prototropy*, which is the movement of an (acidic) hydrogen atom and a double bond.

For most carbonyl compounds, the keto structure is greatly preferred, mainly due to the extra strength of the C=O bond. However, the enol form is stabilised if the C=C bond is *conjugated* with a second π-system or if the OH group is involved in *intramolecular hydrogen bonding*.

Example

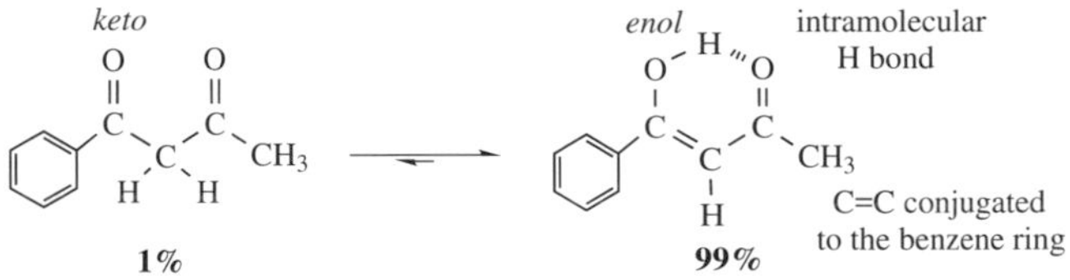

keto

1%

enol

intramolecular H bond

C=C conjugated to the benzene ring

99%

8.4.2   Reactivity of enols

Enols behave as nucleophiles and react with electrophiles at the α-position.

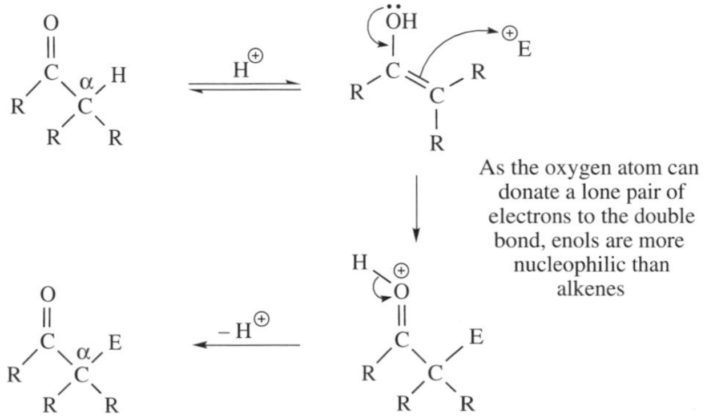

As the oxygen atom can donate a lone pair of electrons to the double bond, enols are more nucleophilic than alkenes

8.4.2.1 α-Halogenation of aldehydes and ketones

Halogenation can be achieved by reaction with chlorine, bromine or iodine in acidic solution.

Example

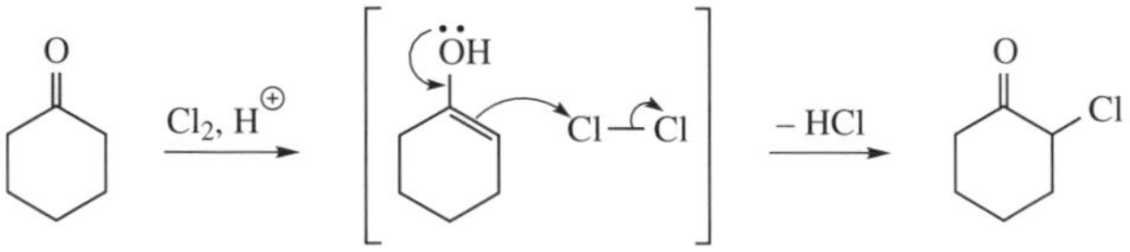

8.4.3 Acidity of α-hydrogen atoms: enolate formation

Bases can abstract α-protons from carbonyl compounds to form enolate anions.

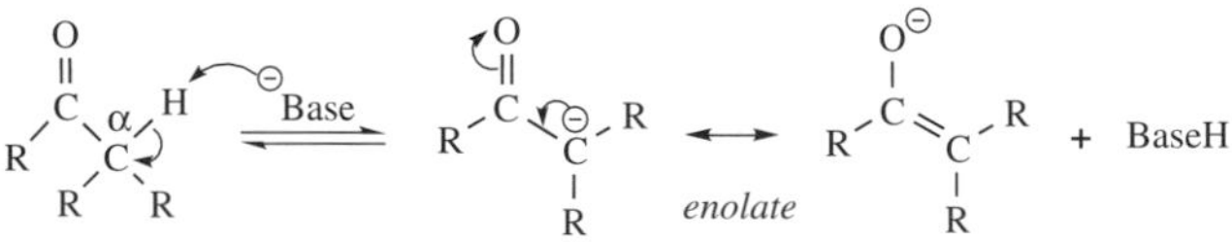

Carbonyl compounds are more acidic than, for example, alkanes because the anion can be stabilised by resonance (see Section 4.3.1).

pK_a	Compound	Anion
60	H_3C—CH_3	H_3C—$\overset{\ominus}{C}H_2$
20	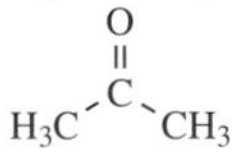 $H_3C\overset{O}{\overset{\|}{C}}CH_3$	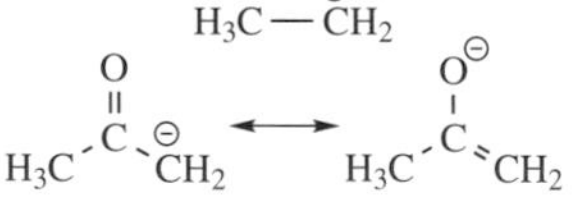 $H_3C\overset{O}{\overset{\|}{C}}\overset{\ominus}{C}H_2 \longleftrightarrow H_3C\overset{O^\ominus}{\overset{\|}{C}}CH_2$
9	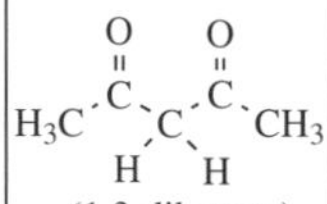 $H_3C\overset{O}{\overset{\|}{C}}\underset{H}{C}\underset{H}{\overset{O}{\overset{\|}{C}}}CH_3$ (1,3-diketone)	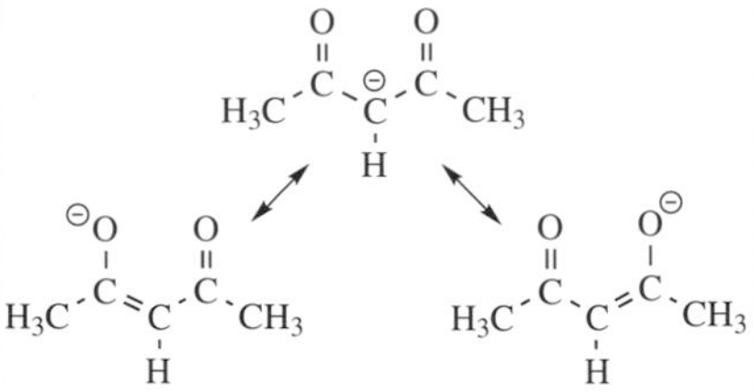

1,3-Diketones (or β-diketones) are therefore more acidic than water (which has a pK_a-value of 16), as the enolate is stabilised by resonance over both carbonyl groups.

8.4.4 Reactivity of enolates

Enolates are much more reactive towards electrophiles than enols because they are negatively charged. Enolates can react with electrophiles on oxygen, although reaction on carbon is more commonly observed.

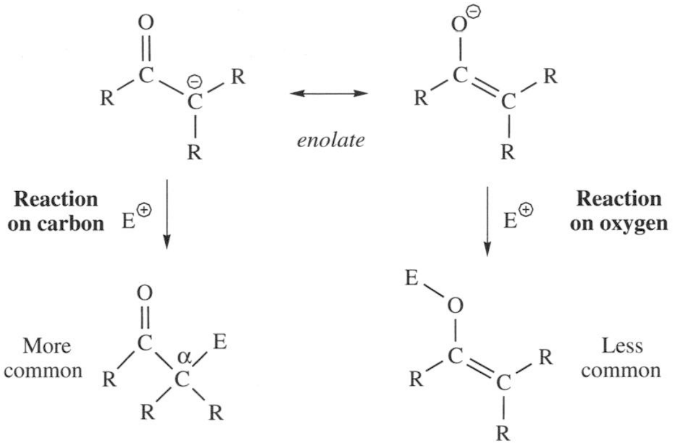

8.4.4.1 Halogenation of enolates

The reaction of methyl ketones with excess hydroxide and chlorine, bromine or iodine leads to the formation of a carboxylic acid together with CHX_3 in a *haloform reaction*. The use of iodine gives CHI_3 (iodoform), which is the basis of a functional group test for methyl ketones.

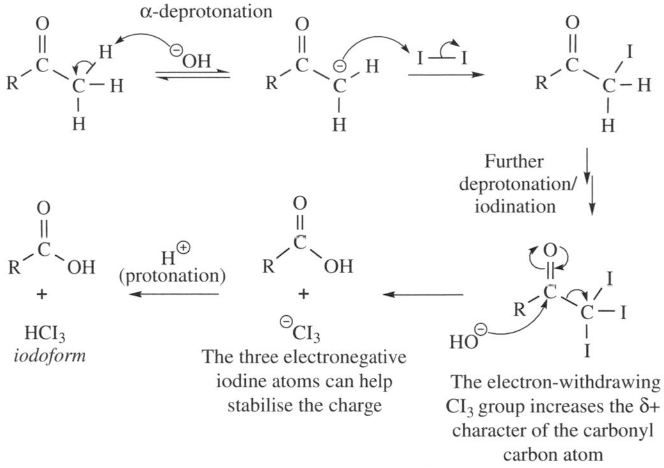

Therefore, although halogenation under acidic conditions leads to the formation of a monohalogenated product, halogenation under basic conditions leads to the formation of a polyhalogenated product.

8.4.4.2 Alkylation of enolate ions

Enolates can be alkylated by reaction with alkyl halides (in an S_N2 reaction with primary and secondary alkyl halides). These reactions produce new carbon–carbon bonds.

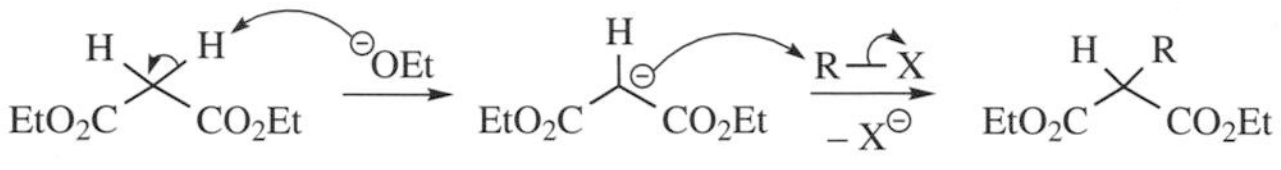

8.5   Carbonyl–carbonyl condensation reactions

These reactions, which involve both nucleophilic addition and α-substitution steps, are amongst the most useful carbon–carbon bond-forming reactions in synthesis.

8.5.1   Condensations of aldehydes and ketones: the aldol reaction

This is a base-catalysed dimerisation reaction for all aldehydes and ketones with α-hydrogens.

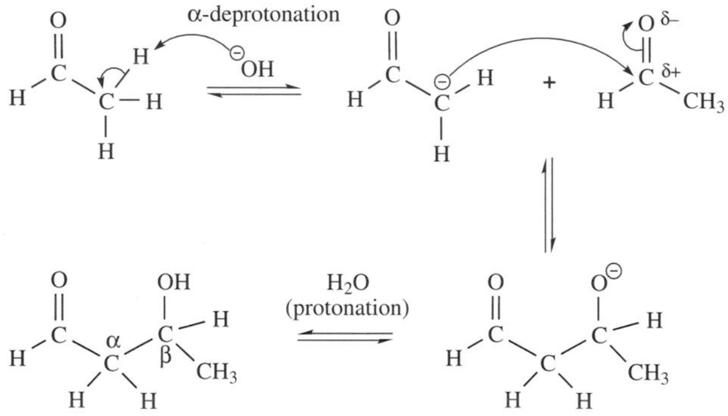

3-hydroxybutanal (aldol)

   With aldehydes, this rapid and reversible reaction leads to the formation of a β-hydroxy aldehyde or *aldol* (*ald* for aldehyde and *ol* for alcohol). When using ketones, β-hydroxy ketones are formed. The aldol products can undergo loss of water on heating under basic or acidic conditions to form conjugated enones (containing both C=O and C=C bonds) in condensation reactions. Conjugation stabilises the enone product and this makes it relatively easy to form.

Base catalysed (E1cB mechanism)

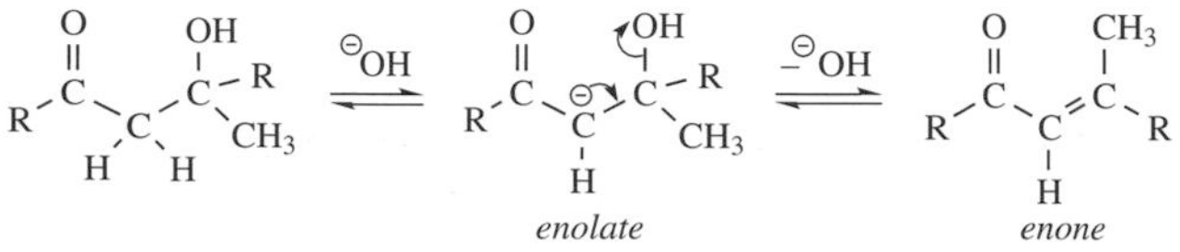

Acid catalysed

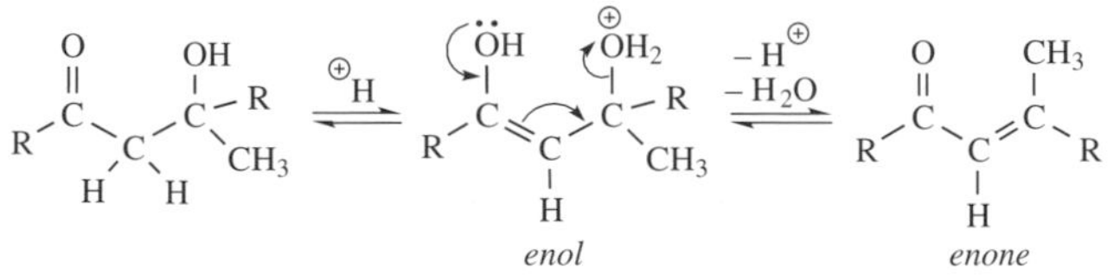

enol   *enone*

8.5.2 Crossed or mixed aldol reactions

These reactions involve two different carbonyl partners. If two similar aldehydes or ketones are reacted, a mixture of all four possible products is formed (e.g. reaction of A and B gives AA, AB, BA and BB). This is because both carbonyl partners can act as nucleophiles and electrophiles.

The formation of only one product requires the following:

(1) Only one carbonyl partner can have α-hydrogens. This means that only one carbonyl partner can be deprotonated to form an enolate nucleophile.

(2) The carbonyl partner without α-hydrogens must be more electrophilic than the carbonyl partner with α-hydrogens.

(3) The carbonyl partner with α-hydrogen atoms should be added slowly. This will ensure that as soon as the enolate is generated it will be trapped by the carbonyl without α-hydrogen atoms (as this is present in high concentration).

Example

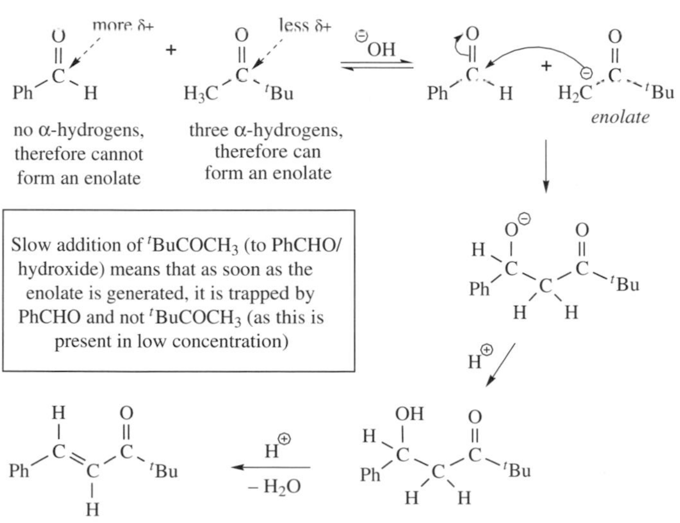

no α-hydrogens, therefore cannot form an enolate

three α-hydrogens, therefore can form an enolate

enolate

Slow addition of $^tBuCOCH_3$ (to PhCHO/ hydroxide) means that as soon as the enolate is generated, it is trapped by PhCHO and not $^tBuCOCH_3$ (as this is present in low concentration)

8.5.3   Intramolecular aldol reactions

Intramolecular cyclisation reactions can occur so as to form stable 5- or 6-membered cyclic enones. These ring sizes are preferred over strained 3- and 4-membered rings (see Section 3.2.3) or difficult-to-make medium-sized (8–13-membered) rings. (Large rings are difficult to make because as the precursor chain length increases, so does the number of conformations that can be adopted. This means that, for long chains, there is a greater loss of entropy on formation of the cyclisation transition state.)

Example

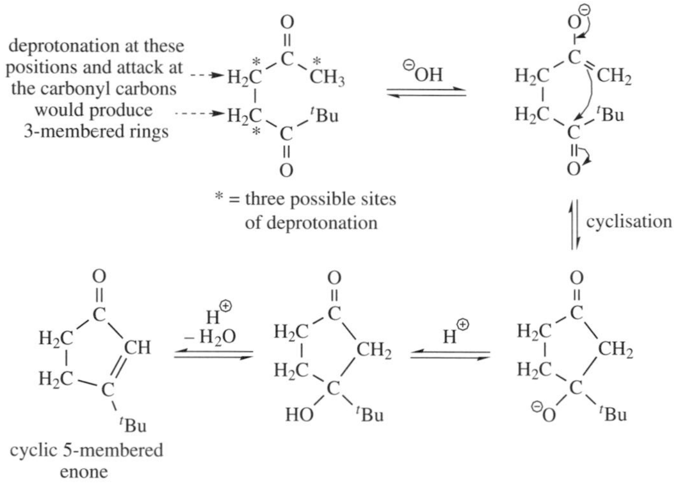

deprotonation at these positions and attack at the carbonyl carbons would produce 3-membered rings

* = three possible sites of deprotonation

cyclisation

cyclic 5-membered enone

8.5.4   The Michael reaction

Enones derived from carbonyl condensations can undergo further carbon–carbon bond-forming reactions on addition of carbon nucleophiles. When the nucleophile adds at the 4-position (rather than the 2-position) of the enone, this is known as the *Michael reaction*.

1,2-Addition

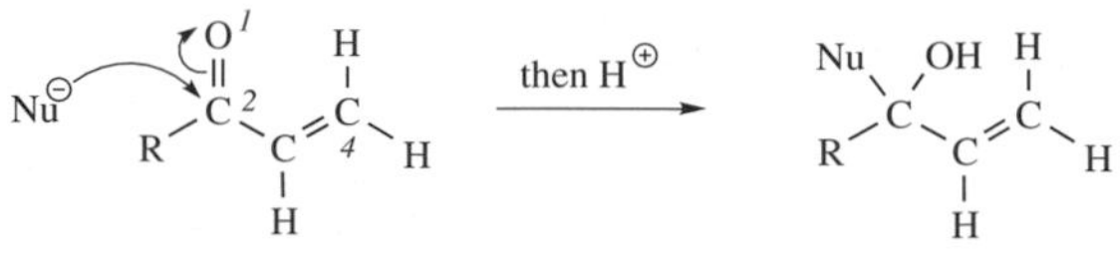

then $H^{\oplus}$

attack at the carbonyl carbon

1,4-Addition

attack at the conjugated alkene

The site of attack can depend on the nature of the nucleophile. Organolithium compounds (RLi) and Grignard reagents (RMgX) tend to give 1,2-addition. For 1,4-addition, organocopper reagents (e.g. R_2CuLi, which are known as cuprates) can be used. The change in the site of attack has been explained by the *hard and soft principle*, as described below.

- *Hard nucleophiles* are those with a negative charge localised on one small atom (e.g. MeLi). These tend to react at the carbonyl carbon, which is known as a *hard electrophilic centre*.
- *Soft nucleophiles* are those in which the negative charge is delocalised (e.g. spread over the large copper atom in R_2CuLi). These tend to react at the 4-position, which is known as a *soft electrophilic centre*.

Problems

(1) The organic product of the reaction of an aldehyde (RCHO) with $NaBH_4$ followed by aqueous acid is RCH_2OH.
 (a) Provide a mechanism for this reaction.
 (b) Suggest a method for the conversion of RCH_2OH back to RCHO.

(2) The organic product from the reaction of methanal with phenylmagnesium bromide, PhMgBr, followed by aqueous acid, is $PhCH_2OH$.
 (a) Provide a mechanism for this reaction.
 (b) Suggest how you would prepare PhMgBr.

(3) Ethanoylbenzene (acetophenone), $PhCOCH_3$, reacts with iodine–potassium iodide solution and aqueous sodium hydroxide to give a yellow solid **A**, relative molecular mass 394, which is removed on filtration. Treatment of the aqueous filtrate with HCl produces a white precipitate of compound **B**, which shows an absorption band at $1700 \, cm^{-1}$ in the infrared spectrum.

Write the structures and outline the mechanism of the formation of compounds **A** and **B**.

(4) Cyclopentanone reacts with ethane-1,2-diol ($HOCH_2CH_2OH$) in the presence of an acid to form a compound **C**, $C_7H_{12}O_2$.
  (a) Give a structure for **C**.
  (b) Explain the role of the acid.
  (c) Provide a mechanism for the reaction.

(5) Reaction of CH_3CHO with aqueous sodium hydroxide produces a new compound **D**, $C_4H_8O_2$. When **D** is heated with dilute acid, compound **E** is formed together with water.
  (a) Give a structure for **D** and provide a mechanism for its formation.
  (b) Give a structure for **E** and provide a mechanism for its formation.

(6) Reaction of 3-methyl-1-butanol with pyridinium chlorochromate (PCC) gives compound **F**. When **F** is treated with ammonium chloride and potassium cyanide, a new compound, **G**, is formed. Heating **G** with aqueous hydrochloric acid then affords a salt **H**, $C_6H_{13}NO_2 \cdot HCl$.

  Give structures for compounds **F–H**.

9. CARBONYL COMPOUNDS: CARBOXYLIC ACIDS AND DERIVATIVES

Key point. Carboxylic acids and carboxylic acid derivatives have an electronegative group (e.g. OH, OR, NHR or halogen) directly bonded to the carbonyl (C=O). These compounds generally undergo *nucleophilic substitution reactions* in which a nucleophile replaces the electronegative group on the carbonyl. The inductive and mesomeric effects of the electronegative group determine the relative reactivity of these compounds towards nucleophilic attack. Carboxylic acid derivatives with an α-hydrogen atom can form enolates (on deprotonation), which can react in *α-substitution reactions* or in *carbonyl–carbonyl condensation reactions*.

9.1 Structure

Carboxylic acids are members of a class of acyl compounds which all contain an electronegative group bonded to RCO.

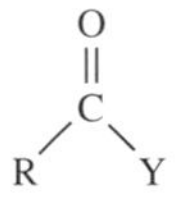

Y	Functional Group
OH	carboxylic acid
halogen	acid halide
OCOR	acid anhydride
OR	ester
NH_2	amide

9.2 Reactivity

As the electronegative group (Y) can act as a leaving group, carboxylic acid derivatives undergo nucleophilic acyl substitution reactions (i.e. reactions leading to the substitution of the Y group by the nucleophile).

With charged nucleophiles

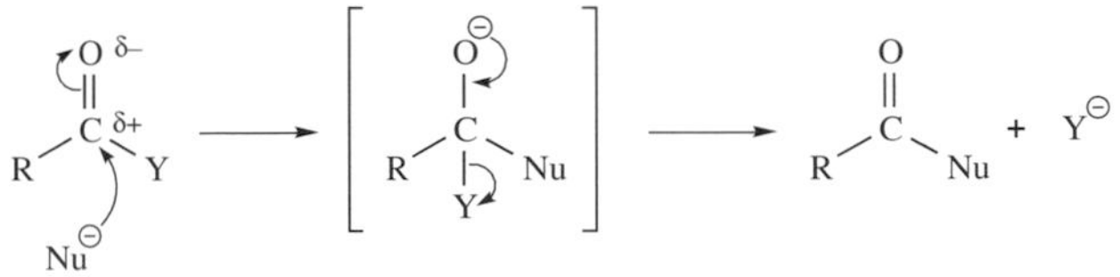

Y = X (halogen), OCOR, OR or NH_2. Nucleophiles = $H^\ominus$, $R^\ominus$

With uncharged nucleophiles

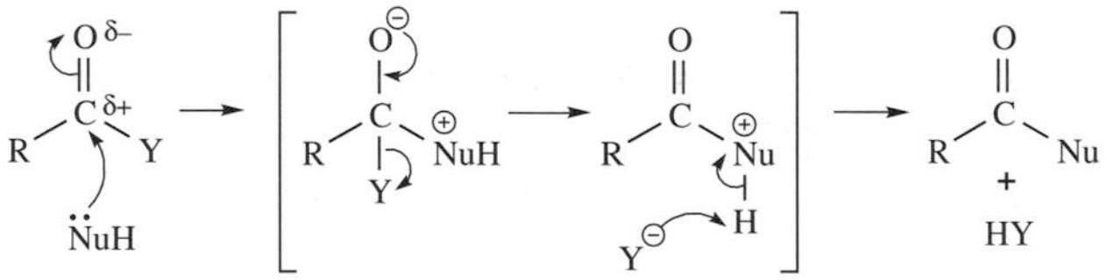

Y = X (halogen), OCOR, OR or NH_2
Nucleophiles = H_2O, ROH, NH_3, RNH_2, R_2NH
This process can be acid catalysed

Carboxylic acid derivatives, like aldehydes and ketones, can also undergo α-substitution and carbonyl–carbonyl condensation reactions.

9.3 Nucleophilic acyl substitution reactions

9.3.1 Relative reactivity of carboxylic acid derivatives

The addition of the nucleophile to the carbonyl carbon is usually the rate-determining step in the substitution reaction. Therefore, the more electropositive the carbon atom (of the acyl group), the more reactive the carboxylic acid derivative.

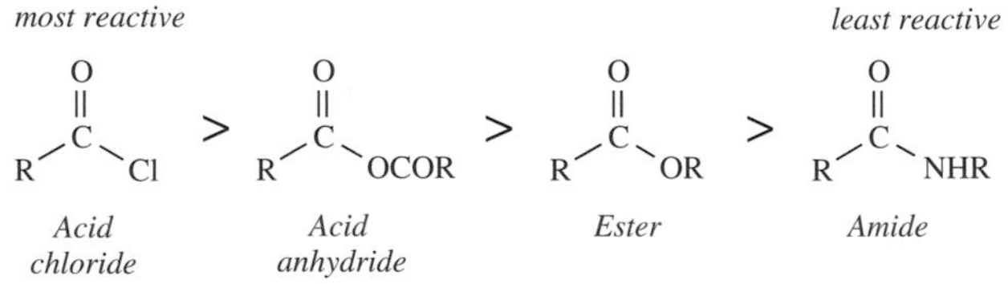

This can be understood from the inductive and mesomeric effects of the electronegative substituent. For example, the +M effect of Cl is much weaker than that of the NH_2 group because the lone pair on chlorine does not interact well with the 2p-orbital on carbon (see Section 7.3.2.2). Therefore, whereas Cl withdraws electrons from the carbonyl group, the NH_2 group mesomerically donates them.

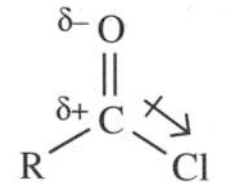

For Cl: –I > +M

The chlorine *withdraws* electrons from the acyl group making the carbonyl carbon very δ+ and hence very susceptible to nucleophilic attack

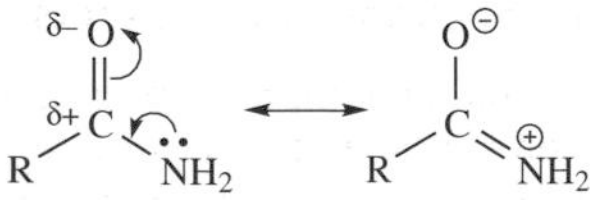

For NH_2: +M > –I

The nitrogen *donates* electrons to the acyl group making the carbonyl carbon less δ+ and hence less susceptible to nucleophilic attack

It is usually possible to transform a more reactive carboxylic acid derivative into a less reactive derivative (e.g. an acid chloride into an amide).

9.3.2 Reactivity of carboxylic acid derivatives versus carboxylic acids

As a carboxylic acid has an acidic hydrogen atom, nucleophiles may act as bases and deprotonate the acid rather than attack the carbonyl carbon atom.

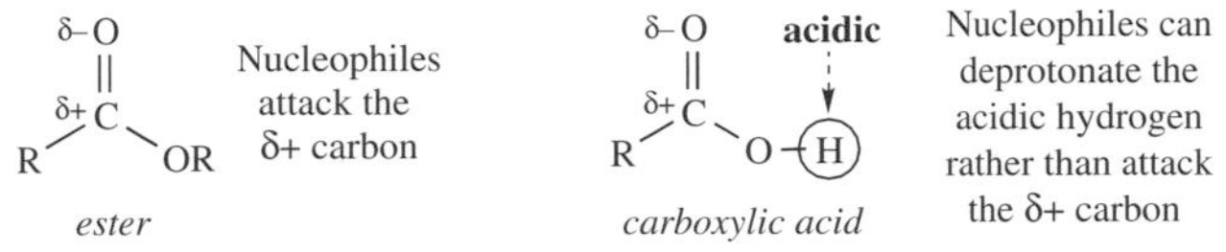

9.3.3 Reactivity of carboxylic acid derivatives versus aldehydes/ketones

Aldehydes and ketones are generally *more* reactive to nucleophiles than esters or amides because the carbonyl carbon atom is more electropositive. In esters, the +M effect of the OR group means that electrons are donated to the acyl group, lowering its reactivity to nucleophiles.

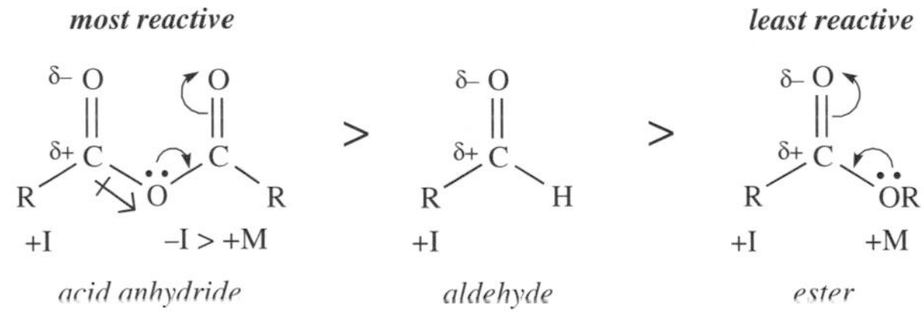

Aldehydes and ketones are generally *less* reactive to nucleophiles than acid halides or acid anhydrides because the carbonyl carbon atom is less electropositive. In acid anhydrides, the carbon is electropositive because the oxygen lone pair is shared between two carbonyl groups, and hence the +M effect is significantly weakened.

9.4 Nucleophilic substitution reactions of carboxylic acids

Substitution of the OH group is difficult because of competing deprotonation. In order for substitution to occur, the OH group needs to be converted into a good leaving group (e.g. Cl).

9.4.1 Preparation of acid chlorides

Carboxylic acids can be converted into acid chlorides using thionyl chloride ($SOCl_2$) or phosphorus trichloride (PCl_3).

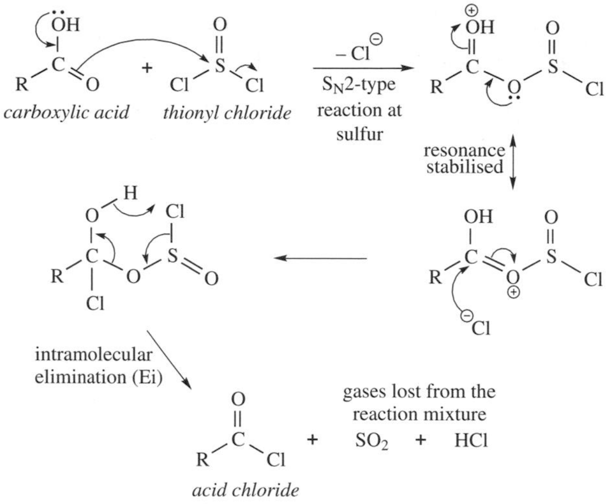

carboxylic acid   thionyl chloride S_N2-type reaction at sulfur

resonance stabilised

intramolecular elimination (Ei)

gases lost from the reaction mixture SO_2 +   HCl

acid chloride

9.4.2   Preparation of esters (esterification)

Carboxylic acids can be converted into esters by reaction with an alcohol (ROH) in the presence of an acid catalyst.

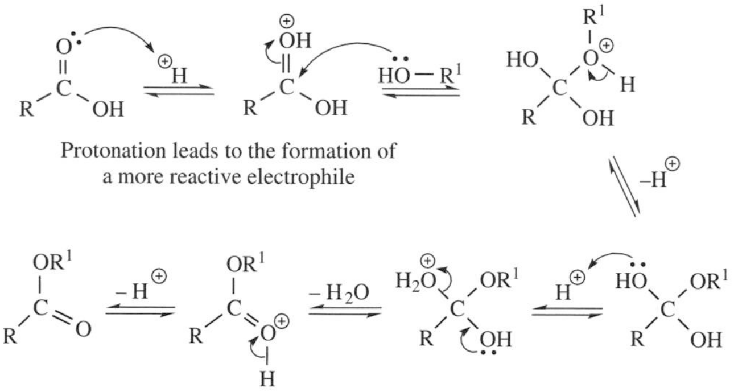

Protonation leads to the formation of a more reactive electrophile

The reaction is reversible, and the formation of the ester usually requires an excess of the alcohol.

9.5   Nucleophilic substitution reactions of acid chlorides

Acid chlorides are very reactive and can be converted into a variety of compounds, including less reactive carboxylic acid derivatives.

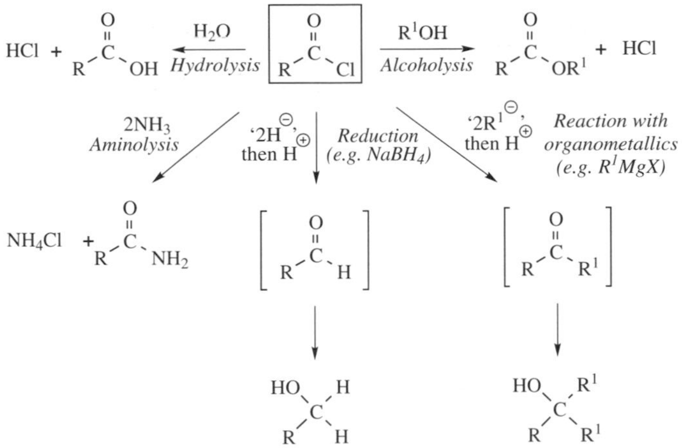

In all cases, nucleophilic acyl substitution leads to the introduction of a nucleophile at the expense of chlorine. Reaction with reducing agents or organometallic compounds can lead to the formation of intermediate aldehydes or ketones. These can subsequently undergo nucleophilic addition reactions (to give alcohols) in the presence of a second equivalent of the reducing agent/organometallic compound.

Example

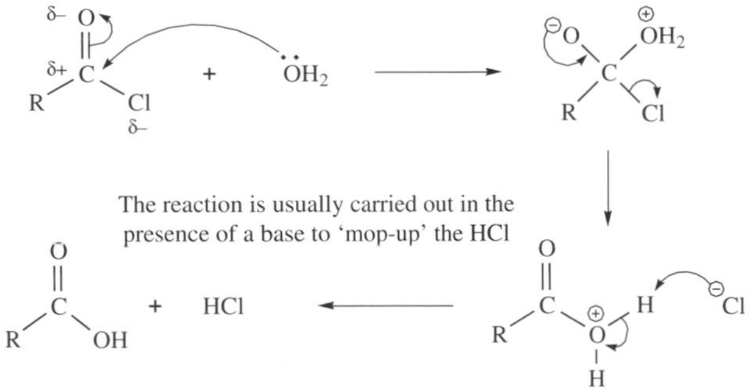

The reaction is usually carried out in the presence of a base to 'mop-up' the HCl

Under the same conditions, carboxylic acids can undergo acid–base reactions to produce salts.

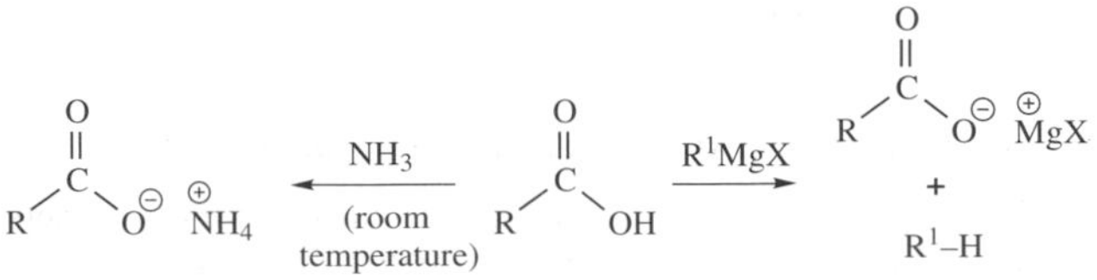

9.6 Nucleophilic substitution reactions of acid anhydrides

Acid anhydrides undergo reactions similar to acid chlorides. These reactions involve the substitution of a carboxylate group (RCO_2^-). Reduction using $LiAlH_4$ can produce an intermediate aldehyde and carboxylate, which are subsequently reduced to alcohols.

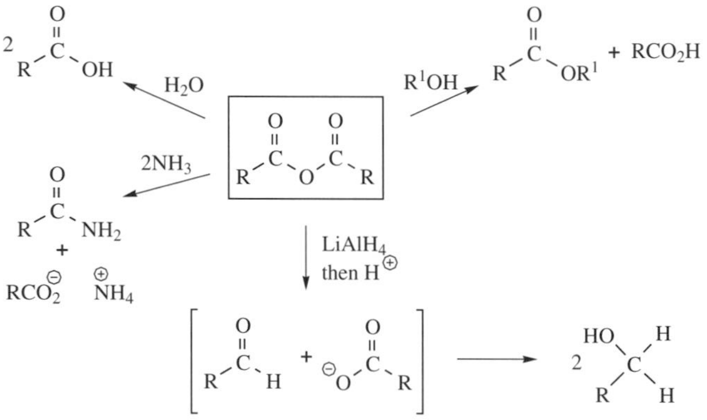

$LiAlH_4$ (but not $NaBH_4$) can reduce carboxylic acids to alcohols by the mechanism shown below.

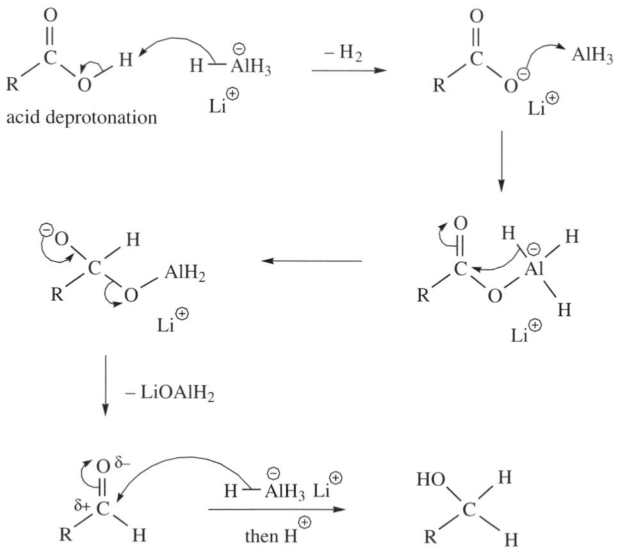

A particularly good reagent for reducing carboxylic acids to alcohols is borane (B_2H_6, which acts as BH_3). For a related reduction, see Section 9.8.

9.7 Nucleophilic substitution reactions of esters

Although less reactive towards nucleophiles than acid chlorides or anhydrides, esters can react with a number of nucleophiles. Reduction usually requires $LiAlH_4$, as $NaBH_4$ reacts very slowly with esters. Indeed, $NaBH_4$ can be used to selectively reduce aldehydes and ketones in the presence of esters.

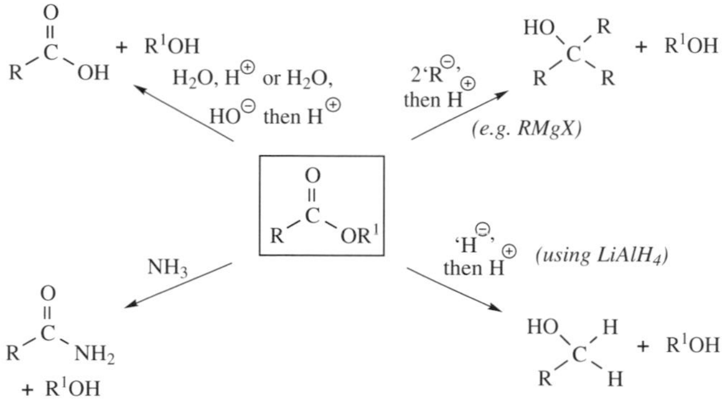

Esters react with two equivalents of Grignard reagents to form tertiary alcohols. This involves a nucleophilic substitution reaction (to form an intermediate ketone), followed by a nucleophilic addition reaction.

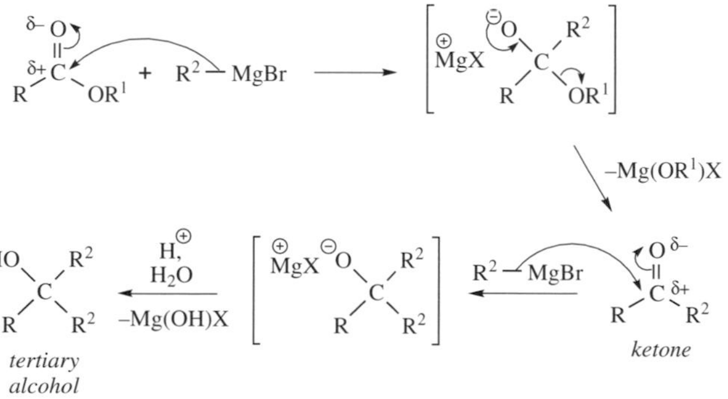

Esters can be hydrolysed in aqueous acid (see Section 9.4.2) or base. In basic solution, this is known as a *saponification* reaction, and this has an important application in the manufacture of soaps from vegetable oils or fats. Whereas acid hydrolysis is reversible, base hydrolysis is *irreversible*, owing to the formation of the resonance-stabilised carboxylate ion.

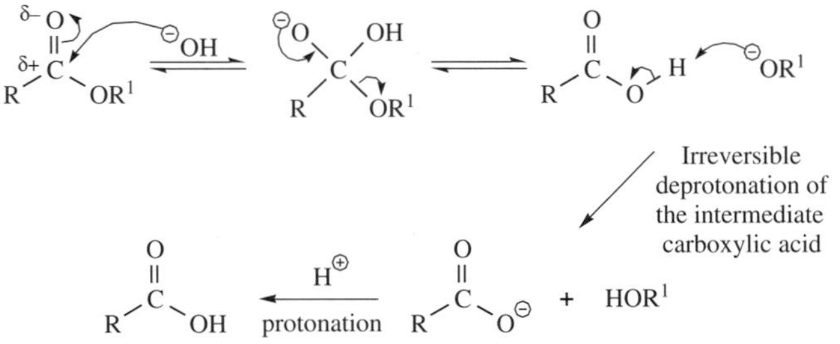

Reaction of an ester with an alcohol under basic or acidic conditions can produce a new ester. This is known as a *transesterification* reaction. The equilibrium can be shifted to the product by removing the alcohol with a low boiling point from the reaction mixture using distillation.

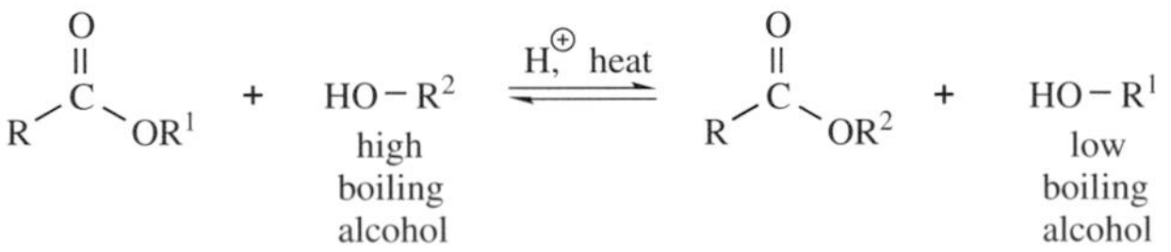

9.8   Nucleophilic substitution and reduction reactions of amides

Amides are much less reactive than acid chlorides, acid anhydrides or esters. Harsh reaction conditions are required for the cleavage of the amide bond, while reduction requires $LiAlH_4$ or borane (B_2H_6, which reacts as BH_3).

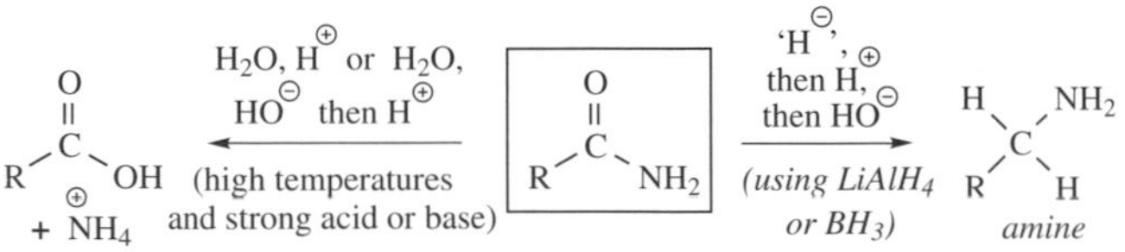

Reduction of primary amides yields primary amines, while secondary and tertiary amides can be reduced to secondary and tertiary amines, respectively. It should be noted that these reactions do not involve nucleophilic substitution of the NH_2 group of the amide.

Lithium aluminium hydride reduction

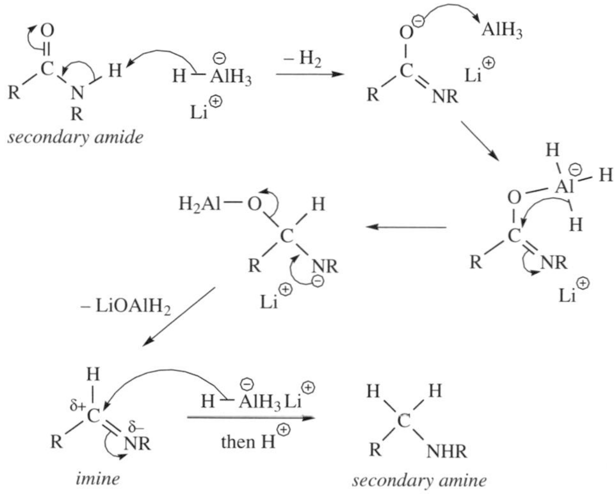

Borane reduction

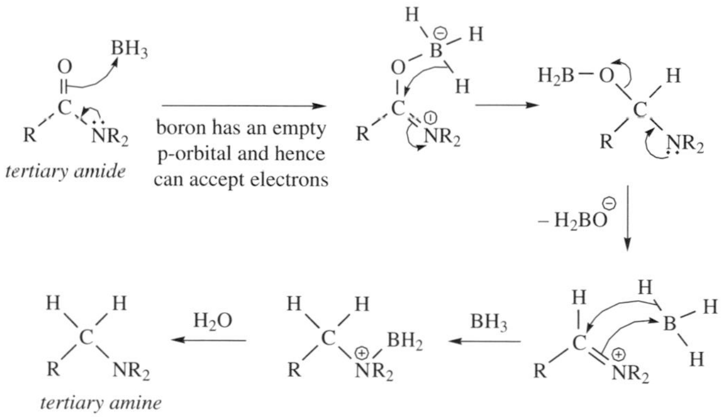

9.9 Nucleophilic addition reactions of nitriles

Nitriles are related to carboxylic acids and derivatives in that the carbon atom of the nitrile is at the same oxidation level as the carbon atom of the acyl group. As a consequence, the reactions of nitriles are similar to that of carbonyls, and nucleophiles attack the nitrile carbon atom.

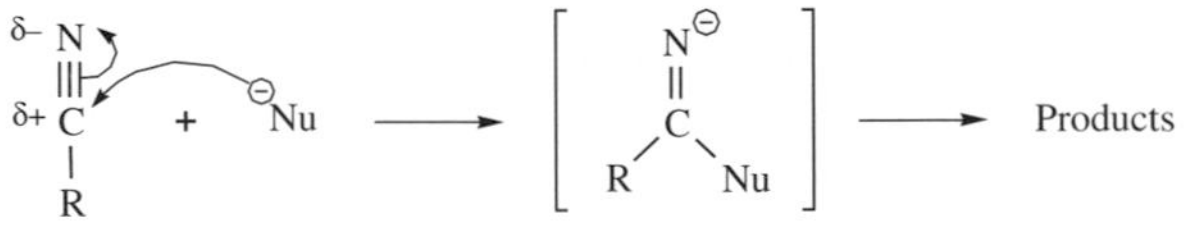

The most important reactions of nitriles are hydrolysis, addition of organometallics and reduction. Reaction with lithium aluminium hydride produces amines, whereas reaction with diisobutylaluminium hydride (DIBAL-H) produces aldehydes. This is because DIBAL-H is a less powerful reducing agent than LiAlH$_4$ (in part because the reagent is more sterically hindered and therefore has more difficulty in transferring hydride ions). So, although reaction with DIBAL-H stops at the imine, reaction with LiAlH$_4$ leads to further reduction of the imine (C=N bond) to give an amine.

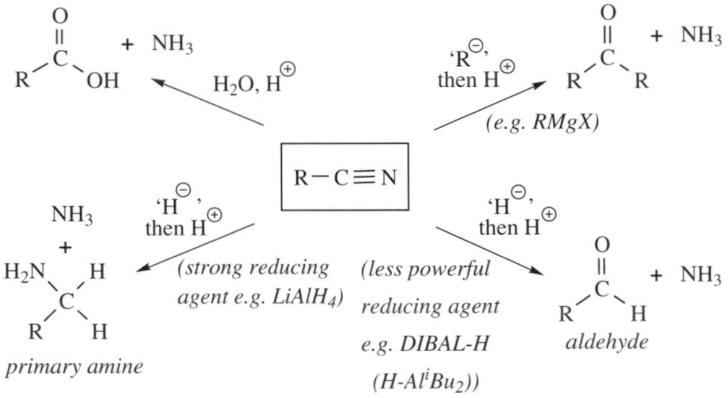

Hydrolysis

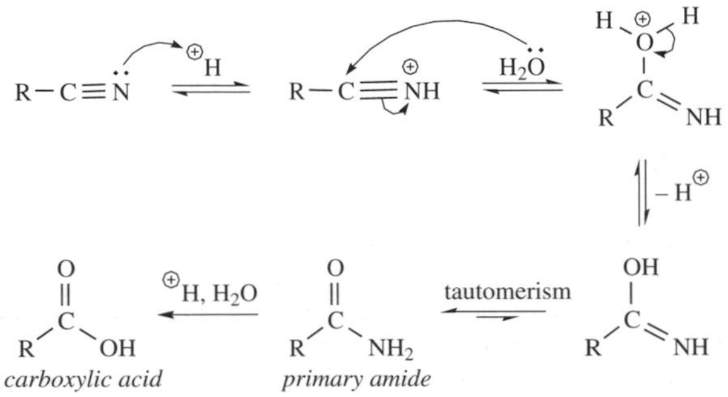

Reduction with DIBAL-H

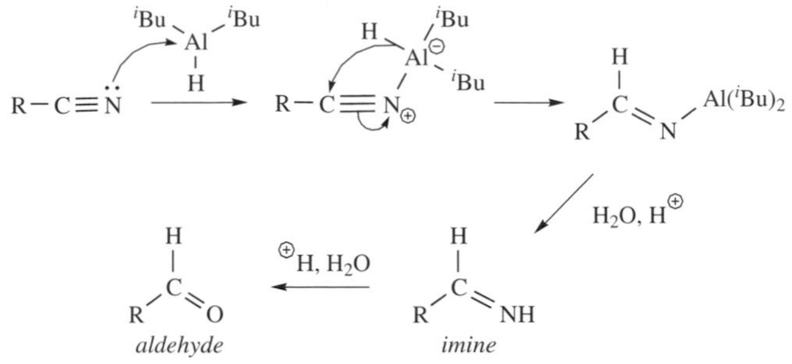

aldehyde   imine

9.10 α-Substitution reactions of carboxylic acids

Carboxylic acids can be brominated at the α-position by reaction with bromine and phosphorus tribromide (PBr$_3$) in the *Hell–Volhard–Zelinsky reaction*. The reaction, which proceeds *via* an acid bromide, leads to the substitution of an α-hydrogen atom by a bromine atom.

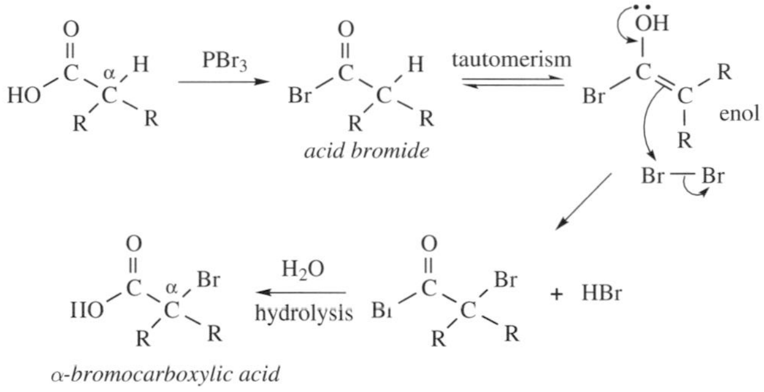

α-bromocarboxylic acid

9.11 Carbonyl–carbonyl condensation reactions

Esters with α-hydrogen atoms can be deprotonated (like aldehydes and ketones) to form resonance-stabilised enolates, which can act as nucleophiles.

9.11.1 The Claisen condensation reaction

A condensation reaction between two esters using one equivalent of base is known as the *Claisen reaction*. One ester loses an α-H atom, while the other loses an alkoxide ion (RO$^-$). The initial steps are reversible, but deprotonation of the intermediate β-keto ester (by the alkoxide ion) shifts the equilibrium to the desired product. It should be noted that the deprotonation of the β-keto ester yields an anion that can be stabilised by delocalisation over two carbonyl groups. At the end of the reaction, acid is added to reprotonate the condensation product.

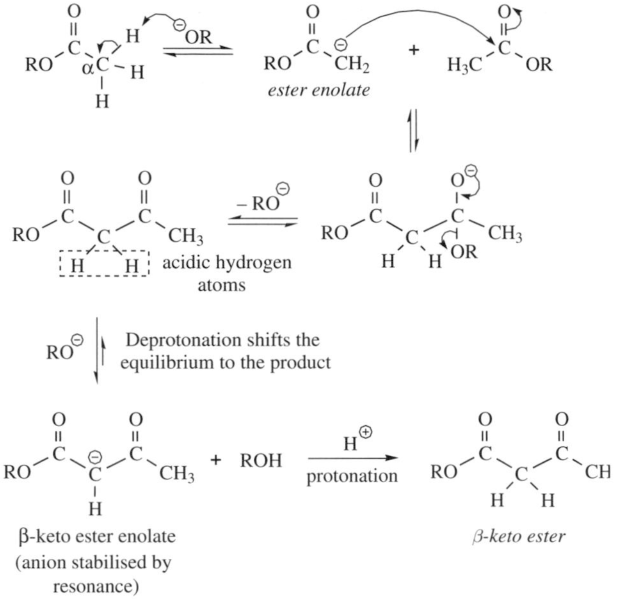

ester enolate

acidic hydrogen atoms

Deprotonation shifts the equilibrium to the product

β-keto ester enolate
(anion stabilised by
resonance)

protonation

β-keto ester

The alkoxide base and ester side chain should be matched. For example, the ethoxide ion should be used as a base for ethyl esters (R = Et in the scheme shown above). This ensures that if the ethoxide ion attacks the carbonyl group of an ethyl ester (in a transesterification reaction; see Section 9.7), then an ethyl ester will be re-formed.

9.11.2   Crossed or mixed Claisen condensations

This involves reaction between two different esters. The reaction works best when osnly one of the esters has α-hydrogen atoms (for the same reasons as for the crossed aldol reaction; see Section 8.5.2).

Example

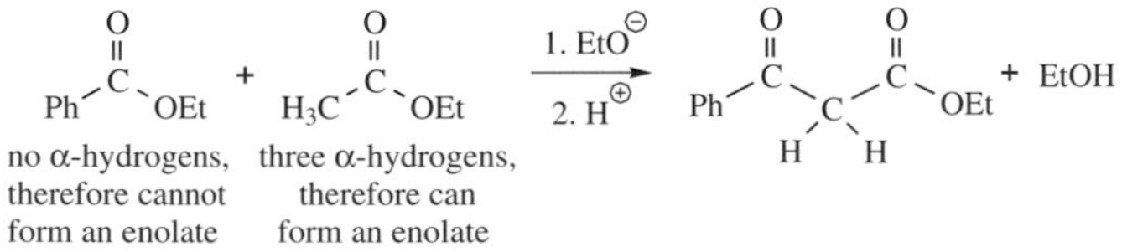

no α-hydrogens,   three α-hydrogens,
therefore cannot   therefore can
form an enolate   form an enolate

Crossed Claisen-like reactions can also occur between esters and ketones. The ester generally acts as the electrophile, as ketones are more acidic than esters (i.e. the ketone enolate, which acts the nucleophile, is more easily formed than an ester enolate). For an ester enolate, the lone pair of electrons on the OR group can compete with the negative charge for delocalisation onto the carbonyl group. This means that the negative charge of an ester enolate is not as readily delocalised onto the carbonyl group as that of a ketone enolate.

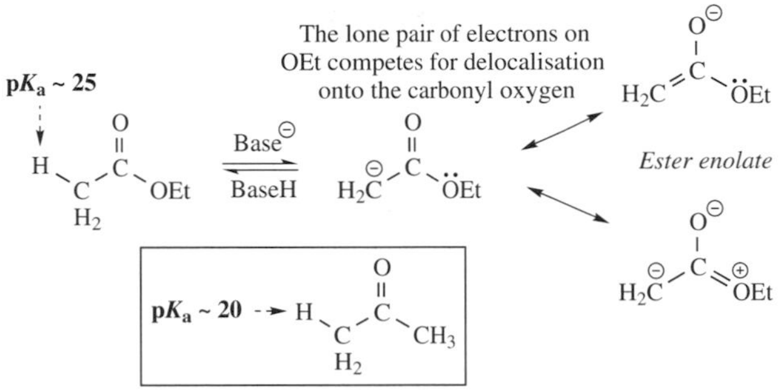

However, the reaction works best when the ester does not contain any α-hydrogen atoms (and hence cannot form an ester enolate).

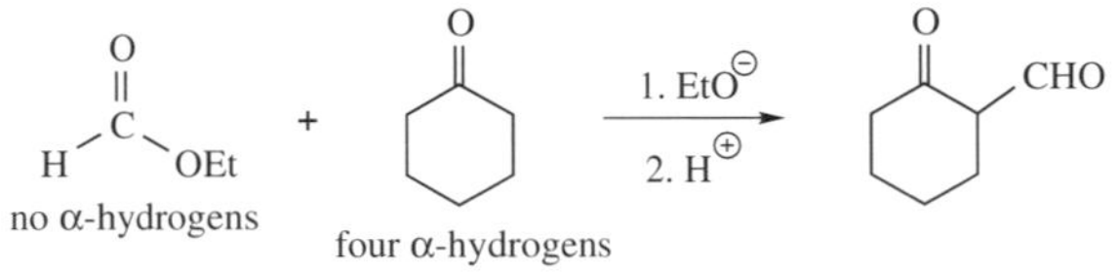

9.11.3 Intramolecular Claisen condensations: the Dieckmann reaction

Intramolecular Claisen reactions are known as *Dieckmann reactions*, and these work well to give 5- or 6-membered rings. A 1,6-diester forms a 5-membered ring, while a 6-membered ring is formed by the condensation of a 1,7-diester.

Example

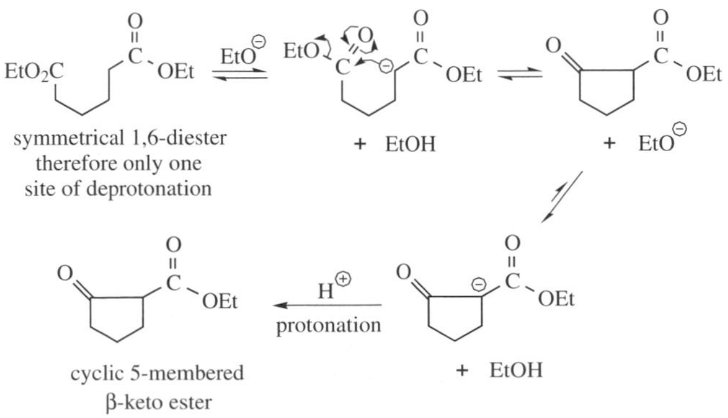

9.12 A summary of carbonyl reactivity

The following guidelines can be used to help predict reaction pathways for aldehydes, ketones (see Chapter 8) and carboxylic acid derivatives.

(1) *Nucleophilic addition versus nucleophilic substitution*

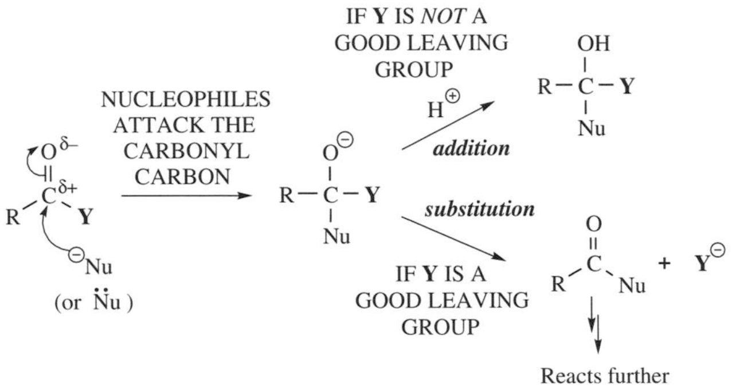

(2) *Nucleophilic addition versus α-deprotonation*

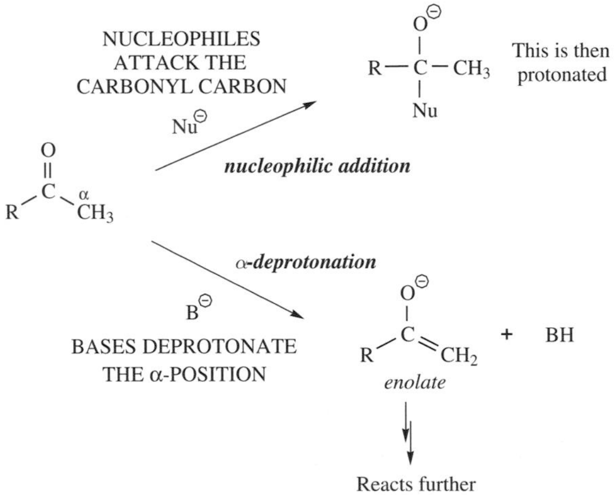

Problems

(1) Lactone **C** can be made from **A** by the following sequence of reactions.

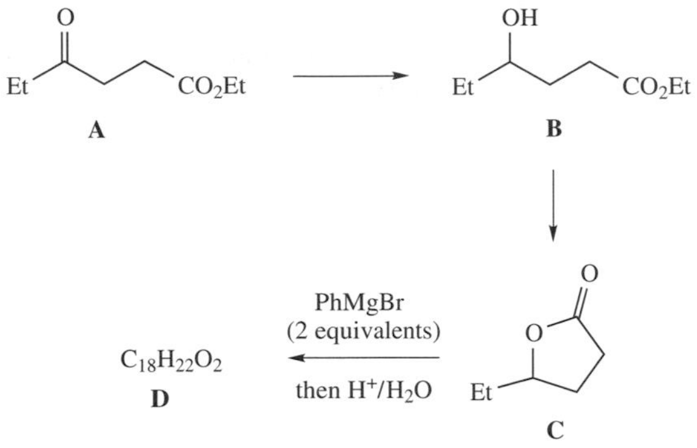

(a) How could **A** be converted into **B**? Explain why the ketone rather than the ester group reacts.

(b) Provide a mechanism to show how **B** can be converted into **C**.

(c) Give the structure of product **D**.

(2) What reagents can be used to accomplish the following transformations?

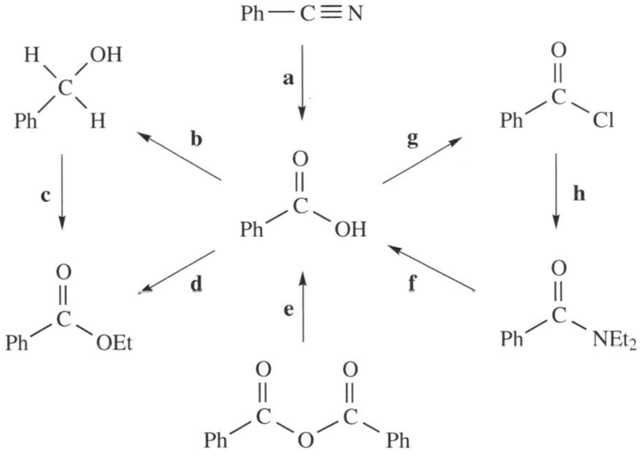

(3) Give the major condensation products from reaction of each of the following starting materials with sodium ethoxide (NaOEt) and then aqueous acid.

(a) $PhCOCH_3 + PhCO_2CH_2CH_3$

(b) $CH_3CH_2OCHO + PhCOCH_3$

(c) $CH_3CO_2CH_2CH_3 + (CH_3)_3CCO_2CH_2CH_3$

(d) $PhCOCH_3 + CH_3CO_2CH_2CH_3$

(4) Heating (S)-2-hydroxybutyric acid with methanol and an acid catalyst gives compound **E**. On reaction of **E** with lithium aluminium hydride followed by water, compound **F** is formed, which reacts with periodic acid (HIO_4) to give **G**. Finally, on reaction of

G with 2,4-dinitrophenylhydrazine (Brady's reagent) and sulfuric acid, a precipitate of compound **H** ($C_9H_{10}N_4O_4$) is formed. Propose the structures for compounds **E–H**.

(5) Propose a mechanism for the following transformation to give compound **I**. (Hint: remember that enones can react with nucleophiles in Michael-type additions.)

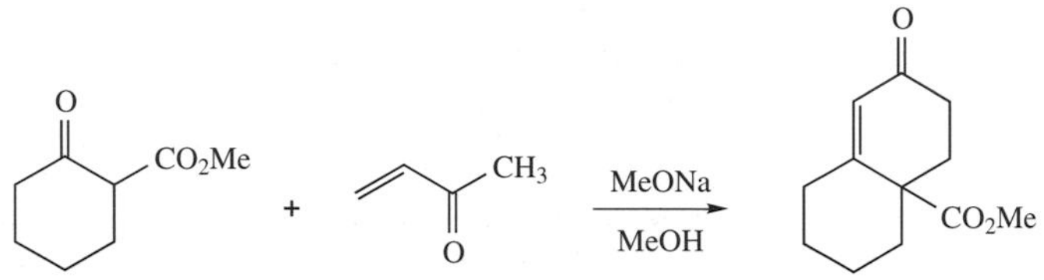

10. SPECTROSCOPY

Key point. Spectroscopy is used to determine the structure of compounds. Spectroscopic techniques include *mass spectrometry* (MS), *ultraviolet* (UV) *spectroscopy, infrared* (IR) *spectroscopy* and *nuclear magnetic resonance* (NMR) *spectroscopy.* MS provides information on the size and formula of compounds by measuring the mass-to-charge ratio of organic ions produced on electron bombardment. UV, IR and NMR spectroscopy rely on the selective absorption of electromagnetic radiation by organic molecules. UV spectroscopy provides information on conjugated π-systems, while IR spectroscopy shows what functional groups are present. The most important method for structure determination is NMR spectroscopy, which can provide information on the arrangement of hydrogen and carbon atoms within an organic molecule.

10.1 Mass spectrometry

10.1.1 Introduction

A mass spectrometer converts organic molecules to positively charged ions, sorts them according to their mass-to-charge ratio (m/z) and determines the relative amounts of the ions present. In electron impact (EI) mass spectrometry (MS), a small sample is introduced into a high-vacuum chamber where it is converted to a vapour and bombarded with high-energy electrons. These bombarding electrons eject an electron from the molecule to give a radical cation, which is called the *parent peak* or *molecular ion.*

$$R\!-\!H \xrightarrow{\ e^{\ominus}\ } \left[R\!-\!H\right]^{\oplus\bullet} + \ e^{\bigcirc}$$

radical cation

The electron that is ejected from the molecule will be of relatively high energy. For example, from a lone pair, which is not involved in bonding.

Examples

$$R\!-\!\overset{..}{N}H_2 \longrightarrow R\!-\!\overset{\oplus\bullet}{N}H_2 + \ e^{\ominus}$$

$$\underset{R}{\overset{\overset{\displaystyle \cdot\overset{..}{O}\cdot}{\|}}{\underset{R}{\diagup}\!C\!\diagdown}} \longrightarrow \underset{R}{\overset{\overset{\displaystyle \cdot\overset{..}{O}\overset{\oplus}{}}{\|}}{\underset{R}{\diagup}\!C\!\diagdown}} + \ e^{\ominus}$$

The molecular ions then pass between the poles of a powerful magnet, which deflects them (the deflection depends on the mass of the ion), before hitting an ion detector. Since the molecular ion has a mass that is essentially identical to the mass of the molecule, mass spectrometers can be used to determine molecular weights.

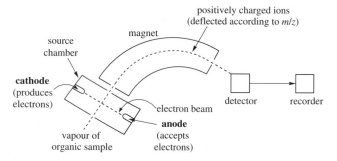

If the bombarding electrons have enough energy, this can lead to the *fragmentation* of the molecular ion into smaller radicals and cations (called daughter ions). Only charged particles (i.e. radical cations and cations) can be recorded by the detector.

$$\left[R\!-\!H \right]^{\oplus \bullet} \longrightarrow \begin{array}{c} H^{\bullet} \ + \ \boxed{R^{\oplus}} \\ or \\ R^{\bullet} \ + \ \boxed{H^{\oplus}} \end{array} \dashleftarrow \begin{array}{l} \text{the cations, but not} \\ \text{the radicals, are} \\ \text{detected} \end{array}$$

radical cation

A mass spectrum determines the masses of the radical cation and cations and their relative concentrations. The most intense peak is called the *base peak*, and this is assigned a value of 100%. The intensities of the other peaks are reported as the percentages of the base peak. The base peak can be the molecular ion peak or a fragment peak.

10.1.2   Isotope patterns

The molecular ion peak (M) is usually the highest mass number, except for the isotope peaks, e.g. $(M+1)$, $(M+2)$, etc. Isotope peaks are present because many molecules contain heavier isotopes than the common isotopes (e.g. ^{13}C rather than ^{12}C).

Example: methane

$\left[CH_4 \right]^{\oplus \bullet}$

peak	M	M+1	M+2
formula	$^{12}C^1H_4$	$^{13}C^1H_4$	$^{13}C^2H^1H_3$
m/z	16	17	18
% of base peak	100	1.14	negligible

The isotope peaks $(M+1)$ and $(M+2)$ are much less intense than the molecular ion peak because of the natural abundance of the isotopes, e.g. $^{12}C = 98.9\%$ and $^{13}C = 1.1\%$. If only C, H, N, O, F, P and I are present, the *approximate* intensities of the (M + 1) and (M + 2) peaks can be calculated as follows.

$$\% \; (M+1) \approx 1.1 \text{ x number of C atoms} + 0.36 \text{ x number of N atoms}$$

$$\% \; (M+2) \approx \frac{(1.1 \text{ x number of C atoms})^2}{200} + 0.20 \text{ x number of O atoms}$$

- The presence of one *chlorine* atom in a molecule can be recognised by the characteristic 3:1 ratio of (M) and (M + 2) peaks in the mass spectrum. This is due to ^{35}Cl (75.8%) and ^{37}Cl (24.2%) isotopes.
- The presence of one *bromine* atom in a molecule can be recognised by the characteristic 1:1 ratio of (M) and (M + 2) peaks in the mass spectrum. This is due to ^{79}Br (50.5%) and ^{81}Br (49.5%) isotopes.

- The presence of *nitrogen* in a molecule can be deduced from the *nitrogen rule*. This states that a molecule of even-numbered molecular weight must contain either no nitrogen or an even number of nitrogen atoms, while a molecule of odd-numbered molecular weight must contain an odd number of nitrogen atoms.

10.1.3   Determination of molecular formula

It is often possible to derive the molecular formula of a compound by recording a high-resolution mass spectrum. As the atomic masses of the isotopes have non-integral masses (except ^{12}C), the measurement of accurate masses to four to five decimal places allows compounds with different atomic compositions to be distinguished.

Example

Compound	$^{12}C^{16}O$	$^{14}N_2$	
Low-resolution *m/z*	28	28	◄---- *same*
	$^{12}C = 12.0000$ $^{16}O = \underline{15.9949}$	$^{14}N = 14.0031$ $^{14}N = \underline{14.0031}$	
High-resolution *m/z*	27.9949	28.0062	◄---- *different*

The top-right of the table also notes: ◄---- *different*

From the molecular formula, the number of sites (or degrees) of unsaturation of an unknown molecule can also be calculated. This number is equal to the sum of the number of rings, the number of double bonds and twice the number of triple bonds. The sites of unsaturation for compounds containing C, H, N, X (halogen), O and S can be calculated from the following formula.

$$\frac{\text{sites of}}{\text{unsaturation}} = \text{carbons} - \frac{\text{hydrogens}}{2} - \frac{\text{halogens}}{2} + \frac{\text{nitrogens}}{2} + 1$$

Example

$$C_7H_7OCl \qquad \frac{\text{sites of}}{\text{unsaturation}} = 7 - \frac{7}{2} - \frac{1}{2} + 1 = 4$$

For example:

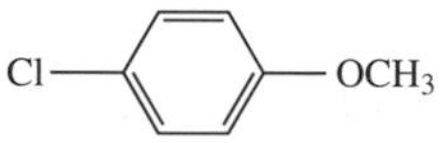

10.1.4 Fragmentation patterns

On EI, molecules first dissociate at the weaker bonds. The resulting fragments may undergo further fragmentation, and analysis of the fragmentation pattern can provide information on the structure of the parent molecule. There are a series of general guidelines, which can be used to predict prominent fragmentation pathways. These rely on the formation of the most stable cation.

(1) Fragmentation of *alkanes* is most likely at highly substituted sites because of the increased stability of tertiary carbocations over secondary and particularly primary carbocations (see Section 4.3).

Example

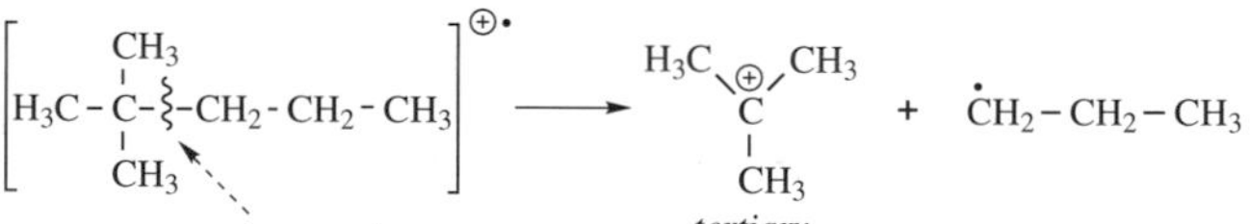

(2) Fragmentation of *alkyl halides* (R–X) usually leads to the cleavage of the weak carbon–halogen bond to form a carbocation (R^+) and a halogen radical (or atom, $X^{\bullet}$).

(3) Fragmentation of *alcohols*, *ethers* or *amines* usually leads to the cleavage of the C–C bond next to the heteroatom to generate a resonance-stabilised carbocation. This is known as α-cleavage.

Example

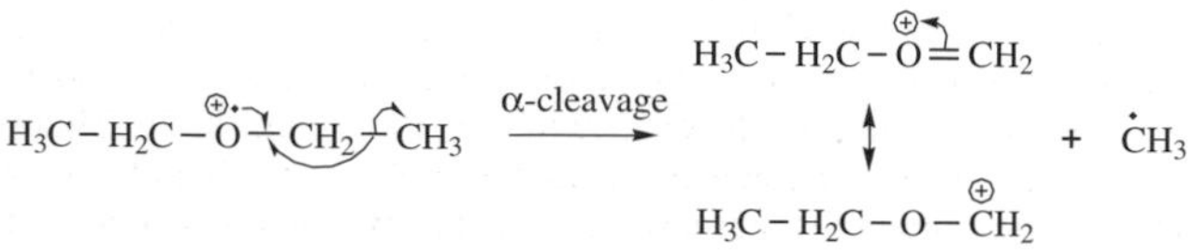

(4) Fragmentation of *carbonyl compounds* usually leads to the cleavage of the C(O)−C bond to generate a resonance-stabilised carbocation. This iss also known as α-cleavage.

Example

10.1.5 Chemical ionisation

Extensive fragmentation can occur in EI spectra because the bombarding electron imparts very high energy to the molecular ion, and fragmentation provides a way to release the energy. A less energetic ionisation technique, which can result in less fragmentation of the molecular ion, is *chemical ionisation* (CI). In this approach, the organic sample is protonated in the gas phase to give a cation (not a radical cation, as for EI), and the $[M+H]^+$ peak is detected by the mass spectrometer (i.e. one mass unit higher than the molecular mass).

10.2 The electromagnetic spectrum

Ultraviolet (UV), infrared (IR) and nuclear magnetic resonance (NMR) spectroscopy involve the interaction of molecules with electromagnetic radiation. When an organic molecule is exposed to electromagnetic energy of different wavelengths, and hence different energies (see the equation below), some of the wavelengths will be absorbed, and this can be recorded in an *absorption spectrum*. It should be noted that high frequencies, large wavenumbers and short wavelengths are associated with high energy.

$$E = h\nu = \frac{hc}{\lambda} = h\bar{\nu}c$$

E = energy (kJ mol^{-1})
h = Planck's constant
ν = frequency (s^{-1} or Hz)
λ = wavelength
c = velocity of light
$\bar{\nu}$ = wavenumber (cm^{-1})

The absorption of radiation depends on the structure of the organic compound and the wavelength of the radiation. Different wavelengths of radiation affect organic compounds in different ways. The absorption of radiation increases the energy of the organic molecule. This can lead to: (i) excitation of electrons from one molecular orbital to another in UV spectroscopy; (ii) increased molecular motions (e.g. vibrations) in IR spectroscopy; or (iii) excitation of a nucleus from a low nuclear spin state to a high spin state in NMR spectroscopy.

Type of spectroscopy	Radiation source	Energy range $(kJ\ mol^{-1})$	Type of transition
Ultraviolet	ultraviolet light	595–298	Electron excitation
Infrared	infrared light	8–50	Molecular vibrations
NMR	radiowaves	25–251×10^{-6}	Nuclear spin

10.3   Ultraviolet spectroscopy

The organic molecule is irradiated with UV light of changing wavelength and the absorption of energy is recorded. The absorptions correspond to the energy required to excite an electron to a higher energy level (e.g. from an occupied orbital to an unoccupied or partially occupied orbital).

Ultraviolet (UV) spectrometer

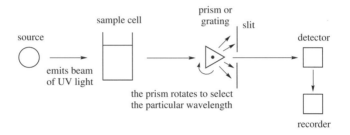

The exact amount of UV light absorbed at a particular wavelength is expressed as the compound's *molar absorptivity* or *molar extinction coefficient* (ε). This is a good estimate of the efficiency of light absorption and is calculated from the absorbance of light, which is derived from the *Beer–Lambert law*.

$$\varepsilon = \frac{A}{c \times l}$$

A = absorbance
c = sample concentration (mol dm^{-3})
l = sample path length (cm)

The Beer–Lambert law: $A = \log_{10}\left(\dfrac{I_0}{I}\right)$

I_0 = intensity of incident light striking sample
I = intensity of transmitted light emerging from sample

The intensity of an absorption band is usually quoted as the molar absorptivity at maximum absorption, ε_{max}. On absorption of UV light, electrons in bonding or non-bonding orbitals (ground state) can be given sufficient energy to transfer to higher energy antibonding orbitals (excited state). The most common excitation of electrons involves $\pi \rightarrow \pi^*$ and $n \rightarrow \pi^*$ transitions.

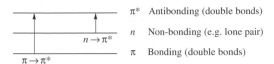

π^* Antibonding (double bonds)

n Non-bonding (e.g. lone pair)

π Bonding (double bonds)

The exact wavelength of radiation required to effect the $\pi \rightarrow \pi^*$ or $n \rightarrow \pi^*$ transition depends on the energy gap, which in turn depends on the nature of the functional group(s) present in the organic molecule. The absorbing group within a molecule, which is usually called the *chromophore*, can therefore be assigned from the wavelength of the absorption peak (λ_{max}).

Chromophore	Transition	λ_{max} (nm)
$-\overset{\cdot\cdot}{\text{O}}-$	$n \rightarrow \pi^*$	≈ 185
C=C	$\pi \rightarrow \pi^*$	≈ 190
C=O:	$n \rightarrow \pi^*$	≈ 300

One of the most important factors which affects λ_{max} is the extent of conjugation (e.g. the number of alternating C–C and C=C bonds). The greater the number of conjugated double bonds, the larger the value of λ_{max}. If a compound has enough double bonds, it will absorb visible light ($\lambda_{max} > 400$ nm) and the compound will be coloured. The UV spectrum can therefore provide information on the nature of any conjugated π-system.

Examples

$\lambda_{max} = 210$ nm ($\pi \rightarrow \pi^*$) $\lambda_{max} = 217$ nm $\lambda_{max} = 258$ nm

An *auxochrome* is a substituent such as NH_2 and OH which, when attached to a chromophore, alters the λ_{max} of and the intensity of the absorption. For example, whereas benzene (C_6H_6) has a λ_{max} of 255 nm, phenol (C_6H_5OH) has a λ_{max} of 270 nm and aniline ($C_6H_5NH_2$) has a λ_{max} of 280 nm. There is a shift to longer wavelengths because the lone pair of electrons on oxygen and nitrogen can interact with the π-system of the benzene ring. Whereas a shift to longer wavelengths is known as a *red shift*, a shift to shorter wavelengths is known as a *blue shift*.

10.4   Infrared spectroscopy

The organic molecule is irradiated continuously with IR light (of changing wavelength), and the absorption of energy is recorded by an

IR spectrometer (which has the same design as a UV spectrometer; see Section 10.3). The absorption corresponds to the energy required to vibrate bonds within a molecule.

The absorption of energy, which gives rise to bands in the IR spectrum, is reported as frequencies, and these are expressed in *wavenumbers* (in cm^{-1}). The most useful region of radiation is between 4000 and 400 cm^{-1}.

$$\text{wavenumber } (\bar{\nu}) = \frac{1}{\text{wavelength } (\lambda)}$$

The frequency of vibration between two atoms depends on the strength of the bond between them and on their atomic weights (*Hooke's law*). A bond can only stretch, bend or vibrate at specific frequencies corresponding to specific energy levels. If the frequencies of the IR light and the bond vibration are the same, then the vibrating bond will absorb energy.

$$\bar{\nu} = \frac{1}{2\pi c}\sqrt{k\frac{(m_1 + m_2)}{m_1 m_2}}$$

$\bar{\nu}$ = vibrational wavenumber (cm^{-1})
k = force constant, indicating the bond strength (N m^{-1})
m_1, m_2 = masses of atoms (kg)
c = velocity of light (cm s^{-1})

The IR spectrum of an organic molecule is complex because all the bonds can stretch and also undergo bending motions. Those vibrations that lead to a change in dipole moment are observed in the IR spectrum. Bending vibrations generally occur at lower frequencies than stretching vibrations of the same group.

symmetric stretch   *symmetric bend (scissoring)*   *symmetric bend (twisting)*

asymmetric stretch   *asymmetric bend (rocking)*   *asymmetric bend (wagging)*

The IR spectrum can therefore be viewed as a unique fingerprint of an organic compound, and the region below 1500 cm^{-1} is known as the *fingerprint region*. Fortunately, the vibrational bands of functional groups in different compounds do not change much, and they appear at characteristic wavenumbers. These bands, particularly stretching vibrations above 1500 cm^{-1}, can provide important structural information.

Bond or functional group	$\bar{\nu}$ (cm^{-1})
RO–H, C–H, N–H	4000–2500
RC≡N, RC≡CR	2500–2000
C=O, C=N, C=C	2000–1500
C–C, C–O, C–N, C–X	<1500

In general, short strong bonds vibrate at higher frequency than long weak bonds (as more energy is required). Therefore, the C≡C bond absorbs at a higher wavenumber than the C=C bond, which in turn absorbs at a higher wavenumber than the C–C bond.

In general, bonds bearing light atoms vibrate at higher frequency than bonds bearing heavier atoms. Therefore, the C–H bond absorbs at a higher wavenumber than the C–C bond.

Alcohols and amines

The intense O–H or N–H stretching vibration between 3650 and 3200 cm^{-1} is the most characteristic peak of alcohols or amines, respectively. If this peak is broad, then this often indicates intermolecular hydrogen bonding.

Carbonyl compounds

The carbonyl functional group exhibits a sharp, intense peak between 1575 and 1825 cm^{-1}, owing to the C=O stretching vibration. The exact position of the peak can be used to identify the type of carbonyl compound.

Carbonyl	$\bar{\nu}$ / cm^{-1}
acid anhydride	1820
acid chloride	1800
ester	1740
aldehyde	1730

Carbonyl	$\bar{\nu}$ / cm^{-1}
ketone	1715
carboxylic acid	1710
amide	1650

10.5 Nuclear magnetic resonance spectroscopy

When certain nuclei of an organic molecule (e.g. ^{1}H, ^{13}C, ^{19}F and ^{31}P) are placed within a strong magnetic field, the nuclear spins align themselves with (parallel to) or against (antiparallel to) the field. Those that are parallel to the field are slightly favoured because they are lower in energy. On irradiation of electromagnetic radiation of the correct frequency, energy can be absorbed to produce *resonance*. This results in 'spin-flipping', and the lower energy nuclei are promoted to the higher energy state. The nuclei can then relax to their original state by releasing energy.

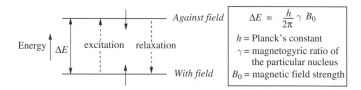

$$\Delta E = \frac{h}{2\pi} \gamma B_0$$

h = Planck's constant
γ = magnetogyric ratio of the particular nucleus
B_0 = magnetic field strength

The energy difference (ΔE) between the two states depends on the magnetic field strength (B_0) and the type of nucleus (γ). These in turn determine the exact amount of radiofrequency energy (ν) which is required for resonance (as $\Delta E = h\nu$). Thus, the smaller the magnetic field, the smaller the energy difference, which means that lower frequency (and lower energy) radiation is needed. The absorption frequency (ν) for ^{1}H and ^{13}C nuclei within the same molecule is not all the same.

Electrons surround the nuclei and produce small local, induced magnetic fields (B_i) that act against the applied magnetic field (B_0). The magnetic field actually felt by the nucleus is therefore a little smaller than the applied field, and the nuclei are said to be *shielded*. Therefore, the more electron density there is near a nucleus, the greater will be the shielding.

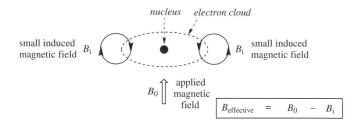

$$B_{effective} = B_0 - B_i$$

- *Electronegative* groups near a nucleus will pull electrons away and *decrease* the shielding of the nucleus.
- *Electropositive* groups near a nucleus will increase the surrounding electron density and *increase* the shielding of the nucleus.

As each type of nucleus has a slightly different electronic environment, each nucleus will be shielded to a slightly different extent. A high-resolution NMR spectrometer can detect the small differences in the effective magnetic fields of the nuclei and produce different NMR peaks for each type of nuclei.

NMR spectrometer

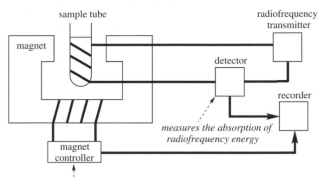

changes the magnetic field strength

The NMR spectrum therefore records the difference in effective field strength (horizontal axis) against the intensity of absorption of radio-frequency energy (vertical axis). In older machines, this is achieved by keeping the radiofrequency constant and varying the strength of the applied magnetic field. The high-field (or upfield) side is on the right, while the low-field (or downfield) side is on the left of the spectrum.

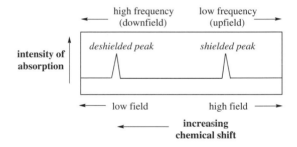

In modern instruments, the magnetic field is kept constant, and the radiofrequency is varied in *pulse Fourier transform NMR* (FT-NMR). In FT-NMR, all of the nuclear spins are excited instantaneously using a mixture of radiofrequencies. The spectrum is obtained by analysing the emission of radiofrequency energy (as the spins return to equilibrium) as a function of time.

The NMR spectrum is calibrated using tetramethylsilane (Me$_4$Si), which produces a single low-frequency peak in both ^{1}H and ^{13}C NMR spectra. This is set a *chemical shift* value of 0 on the delta (δ) scale, where 1 $\delta = 1$ part per million (ppm) of the operating frequency of the spectrometer. Almost all other absorptions occur at higher frequency to this signal, typically 0–10 ppm for ^{1}H NMR and 0–210 ppm for ^{13}C NMR.

$$\delta\,(\text{ppm}) \;=\; \frac{\text{distance of peak from Me}_4\text{Si (Hz)} \;\times\; 10^6}{\text{spectrometer frequency (Hz)}}$$

^{1}H and ^{13}C nuclei are the most commonly observed nuclei, but because they absorb in different radiofrequency regions, they cannot be observed at the *same* time (i.e. in the ^{1}H NMR spectrum, ^{13}C signals are not observed).

10.5.1 ^{1}H NMR spectroscopy

To obtain a ^{1}H NMR (or proton NMR) spectrum, a small amount of sample is usually dissolved in a deuterated solvent (e.g. $CDCl_3$), and this is placed within a powerful magnetic field. The spectrum can provide information on the number of equivalent protons in an organic molecule. Equivalent protons show a single absorption, while non-equivalent protons give rise to separate absorptions. The number of peaks in the spectrum can therefore be used to determine how many different kinds of proton are present. The relative number of hydrogen atoms responsible for the peaks in the ^{1}H NMR spectrum can be determined by *integration* of the peak areas.

10.5.1.1   Chemical shifts

The chemical shift (δ) value of the peak provides information on the magnetic/chemical environment of the protons. Protons next to electron-withdrawing groups are *deshielded* (leading to high δ-values), whereas protons next to electron-donating groups are *shielded* (leading to low δ-values).

Example

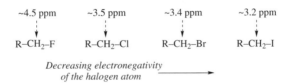

Functional groups therefore have characteristic chemical shift values.

Functional group	δ (ppm)
alkane CH–C	0–1.5
allylic, benzylic CH–C=C	1.5–2.5
alkyl halide CH–X amine CH–NR$_2$ ether CH–OR	2.5–4.5
alkene CH=C	4.5–6.5
aromatic CH=C	6.5–8
aldehyde CH=O	9–10

In general, the chemical shifts of methyl (CH_3) protons appear at low chemical shift values, while that of methylene (CH_2) protons appear at slightly higher values and that of methine (CH) protons at even higher values.

Example

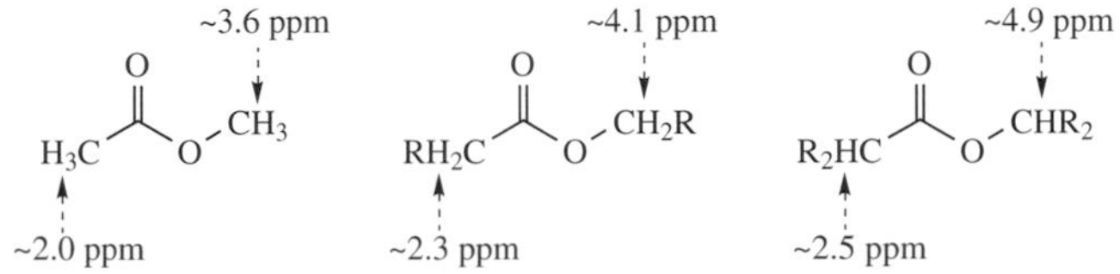

Aromatics, alkenes and aldehydes

Hydrogens bonded to an aromatic ring are strongly deshielded and absorb downfield. When the π-electrons enter the magnetic field, they circulate around the ring to generate a *ring current*. This produces a small, induced magnetic field that reinforces the applied field outside the ring, resulting in the aromatic protons being deshielded. The presence of an aromatic ring current is characteristic of aromatic compounds. A related effect is observed for alkenes and aldehydes. For these compounds, circulation of π-electrons in the double bonds produces induced magnetic fields. These are responsible for the high chemical shift values of alkene protons and, particularly, aldehyde protons.

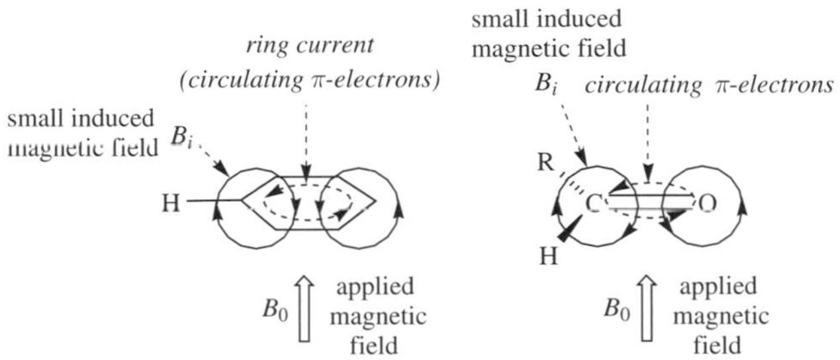

10.5.1.2   Spin–spin splitting or coupling

Peaks in the ^1H NMR spectrum are often not singlets (i.e. single peaks) because of *spin–spin splitting* or *coupling*. The small magnetic field of one nucleus affects the magnetic field of a neighbouring nucleus, and this 'coupling' results in the splitting of the peaks. The distance between the peaks is called the coupling constant (J), which is measured in Hertz (Hz).

The appearance of the peak depends on the number of neighbouring protons. This can be calculated using the $n + 1$ rule, where n is the number of equivalent neighbouring protons. The multiplicity and relative intensities of the peaks can be obtained from Pascal's triangle (shown below).

n	number of peaks ($n+1$)	peak pattern	integration ratios
0	1	singlet (s)	1
1	2	doublet (d)	1 : 1
2	3	triplet (t)	1 : 2 : 1
3	4	quartet (q)	1 : 3 : 3 : 1
4	5	quintet (quin)	1 : 4 : 6 : 4 : 1
5	6	sextet (sex)	1 : 5 : 10 : 10 : 5 : 1
6	7	septet (sep)	1 : 6 : 15 : 20 : 15 : 6 : 1

For spin–spin splitting to occur, the neighbouring hydrogen atoms must be chemically (and magnetically) non-equivalent. One simple way of determining whether hydrogens are equivalent or non-equivalent involves mentally substituting the hydrogens with a phantom group Z. For example, replacement of either of the hydrogen atoms in CH_2Cl_2 with Z will yield the same molecule ($ZCHCl_2$) and hence the hydrogen atoms are equivalent (see below). For $Me(Cl)C=CH_2$, however, replacement of either of the alkene hydrogen atoms with Z will yield alkene diastereoisomers, which are not identical. The two hydrogen atoms are therefore not equivalent and will give different signals in the NMR spectrum.

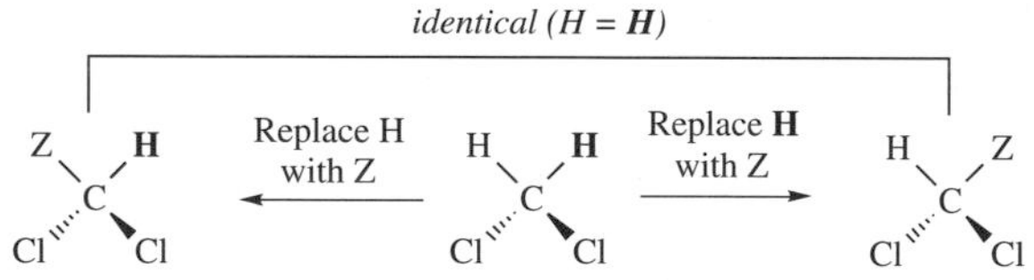

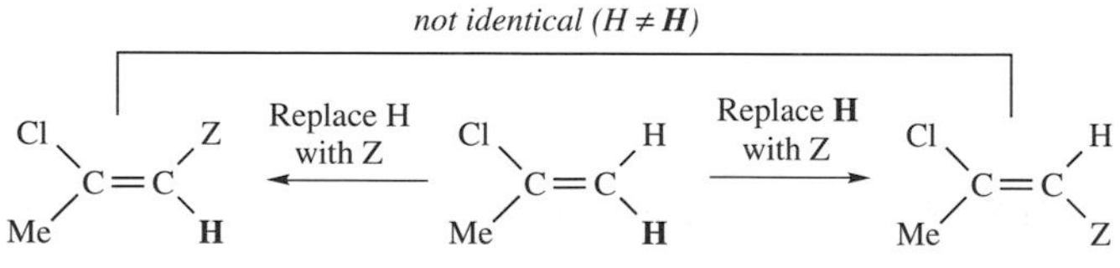

Spin–spin splitting can be observed between non-equivalent hydrogen atoms that are on the same carbon atom (geminal coupling, $J \approx 10$–$20\,Hz$) or adjacent carbon atoms (vicinal coupling, $J \approx 7\,Hz$). Coupling is not usually observed between protons separated by more than three σ-bonds. It should be noted that protons, which are coupled to each other, must have the *same J*-value.

Alkenes

The size of the coupling constant (J) depends on the dihedral angle (ϕ) between hydrogens on adjacent carbon atoms (this is called the Karplus relationship). This can be used to identify cis- and trans-isomers of disubstituted alkenes, RCH=CHR.

- For cis-alkenes, the dihedral angle between the two C–H bonds is 0° (as the two hydrogen atoms are on the same side of the double bond) and $J = 7–11$ Hz.
- For trans-alkenes, the dihedral angle between the two C–H bonds is 180° (as the two hydrogen atoms are on the opposite side of the double bond) and $J = 12–18$ Hz.

Aromatics

For benzene, the splitting between two ortho (1,2-) hydrogen atoms is approximately 8 Hz. For meta (1,3-) and para (1,4-) hydrogen atoms, the splitting is much lower ($J_{meta} = 1–3$ Hz, $J_{para} = 0 –1$ Hz) because the atoms are further apart, and these are known as long-range couplings. Different types of proton often produce signals that overlap to give multiplets (m). This is particularly true for protons of monosubstituted benzenes (C_6H_5–R), which are generally observed as a single overlapping (broad) absorption.

Alcohols

The OH proton of alcohols is generally observed as a single (sharp or broad) peak even when a vicinal proton (CHOH) is present. This is because the alcohol OH protons undergo fast proton exchange with each other (or with traces of water in the sample). This averages the local field, leading to decoupling. Indeed, the peak due to the OH group of the alcohol can be removed in the ^{1}H NMR spectrum by simply shaking a solution of the alcohol with D_2O (to form ROD and HOD). The fact that the peak for ROH disappears and a new peak at ~4.7 ppm arises for HOD can be used to show the presence of an OH (or NH) group.

10.5.1.3 Summary

The ^{1}H NMR spectrum can provide the following structural information:

(1) the types of chemically different protons (from the number of peaks and the chemical shift values);
(2) the number of each type of proton (from integration of the peak areas);
(3) the number of nearest (proton) neighbours that each proton has (from the appearance of the peak, which is determined by spin–spin splitting).

Example

ethyl (*E*)-cinnamate

proton(s)	δ/ppm	peak appearance
H_A	7.64*	doublet (*J* = 16 Hz)
H_B	7.30–7.14	broad multiplet
H_C	6.39	doublet (*J* = 16 Hz)
H_D	4.19	quartet (*J* = 7 Hz)
H_E	1.30	triplet (*J* = 7 Hz)

*This alkene proton is deshielded
because of conjugation with the
(electron-withdrawing) carbonyl group

10.5.2 ^{13}C NMR spectroscopy

The ^{13}C NMR spectrum can provide information on the number of different types of carbon atoms in an organic molecule. Carbons, which are equivalent, show a single absorption, while non-equivalent carbons exhibit separate absorptions.

As ^{13}C has a natural abundance of only 1.1%, the signals are around 1/6000 times as strong as those for ^{1}H. However, spectra can be routinely recorded by FT-NMR using short pulses of radiofrequency radiation. *Broad-band proton decoupling* is usually employed, which removes all ^{13}C–^{1}H coupling, so that singlet peaks are observed for each different kinds of carbon. (As most ^{13}C are surrounded by ^{12}C, ^{13}C–^{13}C coupling is minimal.)

The chemical shift (δ) values of the peaks provide information on the magnetic/chemical environment of the carbons. As for ^{1}H NMR, carbons next to electron-withdrawing groups produce high-frequency signals (leading to high δ-values), whereas carbons next to electron-donating groups produce low-frequency signals (leading to low δ-values).

Functional groups have characteristic carbon chemical shifts, and these generally follow the same trends as for proton chemical shifts. Hence, the order of chemical shifts is usually tertiary (CH) > secondary (CH$_2$) > primary (CH$_3$), and sp^3 carbon atoms generally absorb between 0 and 90 ppm, while sp^2 carbon atoms absorb between 100 and 210 ppm. The chemical shift range is therefore much larger than that for proton NMR, and hence ^{13}C peaks are less likely to overlap.

Alkane carbons	δ (ppm)
CH$_3$	5–20
CH$_2$	20–30
CH	30–50
C	30–45

Functional group	δ (ppm)
alkyl halide C–X	0–80
alcohol/ether C–O	40–80
alkene C=	100–150
aromatic C=	110–160
carbonyl C=O	160–210

Unless special techniques are used, the area under a ^{13}C signal is *not* proportional to the number of carbon atoms giving rise to the signal.

Carbonyls

The carbon atom of the carbonyl group is strongly deshielded and has a characteristic chemical shift of around 160–210 ppm. This chemical shift is higher than that for any other type of carbon atom.

The *DEPT ^{13}C NMR spectrum* (distortionless enhanced polarisation transfer) identifies whether the carbon atom is primary (CH_3), secondary (CH_2), tertiary (CH) or quaternary (C).

The ^{13}C NMR spectrum can therefore provide the following structural information:

(1) the types of chemically different carbons (from the number of peaks and the chemical shift values);
(2) the number of hydrogens on each carbon (when using the DEPT technique).

Example

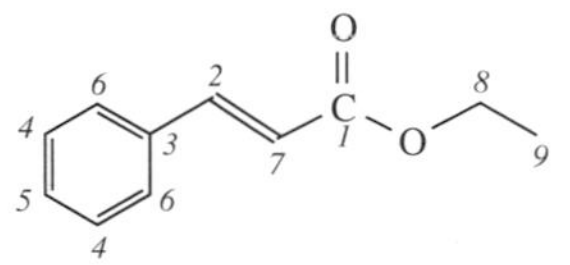

ethyl (*E*)-cinnamate

carbon(s)	peak δ (ppm)
1	165.0
2	142.8
3	134.9
4	128.4
5	127.7
6	126.2
7	117.6
8	59.6
9	13.7

Problems

(1) Compound **A** is believed to have the following structure.

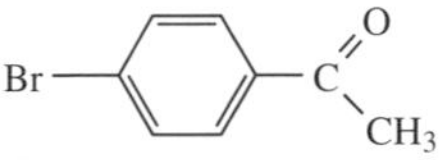

A

Explain how you would use NMR spectroscopy and/or MS to:
(a) confirm the molecular formula;
(b) demonstrate the presence of a single bromine atom;
(c) establish the presence of the $COCH_3$ group;
(d) confirm the presence of a 1,4-disubstituted ring.

(2) Propose a structure for compound **B** using the following spectroscopic information.

m/z (EI) 134 (25%), 91 (85%), 43 (100%). (The percentages refer to the peak heights.)

δ_H 7.40–7.05 (5H, broad multiplet), 3.50 (2H, s), 2.00 (3H, s). (CDCl$_3$ was used as the solvent.)
δ_C 206.0 (C), 134.3 (C), 129.3 (2×CH), 128.6 (2×CH), 126.9 (CH), 48.8 (CH$_2$), 24.4 (CH$_3$). (CDCl$_3$ was used as the solvent.)

(3) Draw the ^{1}H and ^{13}C NMR (fully proton decoupled) spectra for compounds **C** and **D**, indicating the approximate chemical shifts of the protons and carbons and showing spin–spin splitting patterns in the ^{1}H NMR spectra.

C   D

(4) Compound **E** has a prominent IR absorption at 1730 cm^{-1} and gives two singlet peaks in the ^{1}H NMR spectrum. On reaction with sodium borohydride followed by aqueous acid, compound **F** is obtained. Compound **F**, which has a broad IR absorption at 3500–3200 cm^{-1}, reacts with sodium hydride (a base) followed by methyl iodide to give compound **G**. The mass spectrum of **G** shows a molecular ion peak at *m/z* 102 and an intense peak at *m/z* 45. Propose the structures for compounds **E–G**.

11. NATURAL PRODUCTS AND SYNTHETIC POLYMERS

Key point. Natural products are divided into particular classes of compound. These include *carbohydrates, lipids, amino acids, peptides* and *proteins*, and *nucleic acids*. Carbohydrates (or sugars) are polyhydroxycarbonyl compounds that exist as monomers, dimers or polymers. Organic-soluble waxes, oils, fats and steroids are known as lipids, while the condensation of amino acids produces natural polyamides known as peptides and proteins. Nucleic acids (which include RNA and DNA) are natural polymers made up of sugars, heterocyclic nitrogen bases and phosphate groups. Synthetic (unnatural) polymers, which have had a tremendous impact on our day-to-day living, can be classified as chain-growth polymers (e.g. polyethylene) or step-growth polymers (e.g. polyamides or nylons).

11.1 Carbohydrates

Carbohydrates are a class of naturally occurring polyhydroxylated aldehydes and ketones, which are commonly called sugars. Many, but not all, sugars have the empirical formula $C_x(H_2O)_y$.

- *Simple sugars*, or *monosaccharides*, are carbohydrates (such as glucose) which cannot be hydrolysed into smaller molecules.
- *Complex sugars* are carbohydrates, which are composed of two or more sugars joined together by oxygen bridges. These can be hydrolysed to their component sugars (e.g. sucrose is a disaccharide which can be hydrolysed to one glucose molecule and one fructose molecule, while cellulose is a polysaccharide which can be hydrolysed to give around 3000 glucose molecules).

Monosaccharides bearing an aldehyde are known as *aldoses*, whereas those bearing a ketone are known as *ketoses*. Sugars bearing three, four, five and six carbon atoms are known as trioses, tetroses, pentoses and hexoses, respectively.

D-glucose (an aldohexose) — Fischer projection

D-fructose (a ketohexose)

Almost all carbohydrates are chiral and optically active. They are most often drawn as Fischer projections – the horizontal lines come out of the page, while the vertical lines go into the page (see Section 3.3.2.6). If the chiral centre furthest from the aldehyde or ketone is the same absolute configuration as D-(+)-glyceraldehyde, then these are called D-sugars. (Almost all naturally occurring sugars are D-sugars.) Those with the opposite configuration are called L-sugars.

Many carbohydrates exist in equilibrium between open chain and 5- or 6-membered cyclic forms. The cyclic pyranoses (5-ring) or hexanoses (6-ring) are produced from an intramolecular cyclisation of an alcohol group onto an aldehyde, leading to the formation of a hemiacetal (see Section 8.3.5). These are most commonly drawn as *Haworth projections*, in which the ring is drawn as a hexagon and vertical lines attach substituents.

Example: D-glucose

Haworth projections

α-D-glucopyranose   β-D-glucopyranose

The cyclisation produces two diastereoisomers, and the new chiral centre (labelled*), which is formed, is called the *anomeric carbon*. If the anomeric carbon has an *S* configuration, it is called the α-anomer, while the one with an *R* configuration is called the β-anomer.

When α-D-glucopyranose is dissolved in water, the optical rotation value decreases with time until it reaches +52.7°. This is because of equilibration between the α- and β-anomers in a process known as *mutarotation*. The β-anomer undergoes reversible ring opening and re-closure, leading to the formation of some of the α-anomer.

Disaccharides contain two monosaccharide units joined together by a *glycoside linkage*. This linkage occurs between the anomeric carbon of one sugar and a hydroxyl group at any position of another sugar. The bridging oxygen atom is part of an acetal, and hence reaction with aqueous acid (or an appropriate enzyme) leads to hydrolysis and the formation of the component sugars.

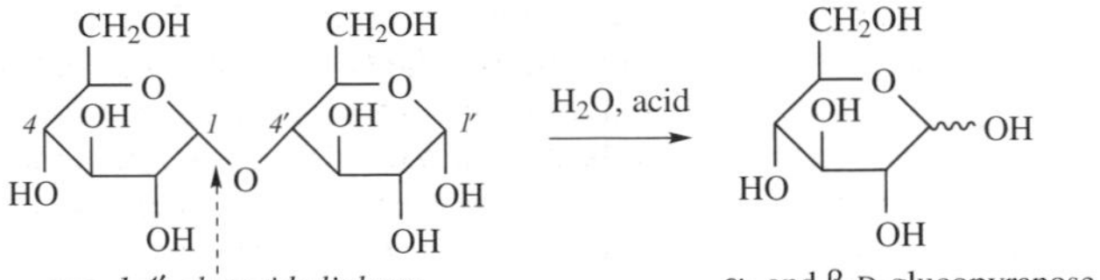

an α-1,4'-glycoside linkage

Maltose
(formed from two D-glucopyranoses)

α- and β-D-glucopyranose

Polysaccharides are polymers of monosaccharides joined by glycoside linkages. The three most abundant natural polysaccharides are cellulose, starch and glycogen, and these are derived from glucose.

11.2 Lipids

Lipids are naturally occurring organic molecules found in cells and tissues, which are soluble in organic solvents but insoluble in water. They are defined by their (organic) solubility rather than their structure. Lipids can be subdivided into fats and waxes, and steroids.

11.2.1 Waxes, fats and oils

Natural waxes are esters of long-chain carboxylic acids with long-chain alcohols. The acid and alcohol residues are usually saturated and contain an even number of carbon atoms.

Example: beeswax

$$H_3C-\left(H_2C\right)_x-\overset{O}{\overset{\|}{C}}-O-\left(CH_2\right)_y-CH_3 \qquad x = 24, 26; \ y = 29, 31$$

Animal fats are solids (e.g. butter), whereas vegetable oils are liquids (e.g. olive oil). They are both triesters of glycerol and are known as *triglycerides*. On hydrolysis, they are converted to glycerol and three carboxylic acids, which are known as fatty acids.

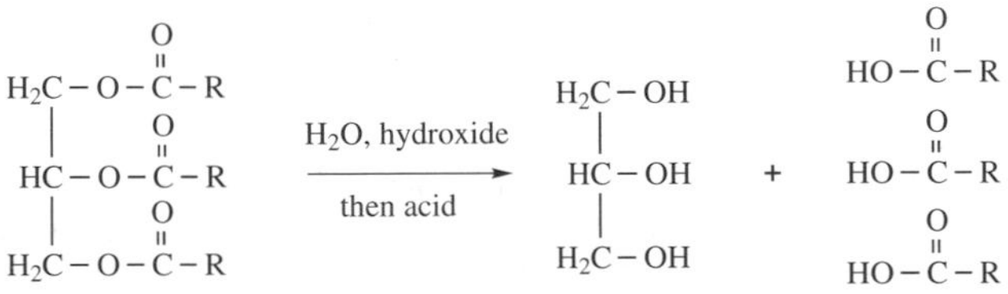

fat or oil
(triester)

glycerol

fatty acids

The fatty acids usually contain between 12 and 20 carbon atoms, and the alkyl side chains can be saturated or unsaturated (i.e. can contain alkene double bonds). Vegetable oils contain a higher proportion of unsaturated fatty acids than animal fats. The shape of the alkenes (which are usually *cis-*) prevents the molecules from packing closely together. This makes it harder for the molecules to crystallise, which lowers the melting point. This explains why (unsaturated) vegetable oils are liquids.

- Catalytic hydrogenation of the alkene double bonds in vegetable oils (known as hardening) is carried out industrially to produce margarine.
- Soap, which is a mixture of sodium and potassium salts of fatty acids, is produced industrially by hydrolysis (saponification) of animal fat using aqueous sodium hydroxide.

11.2.2   Steroids

Naturally occurring steroids exert a variety of physiological activities and many act as hormones (i.e. chemical messengers secreted by glands in the body). They all contain one cyclopentane and three cyclohexane rings linked together: the four rings are labelled A–D.

Example

testosterone
(male sex hormone)

All three cyclohexane rings can adopt strain-free chair conformations, in which the small groups (e.g. hydrogen atoms) at the ring junctions adopt axial positions. Therefore, most steroids have the 'all *trans-*' stereochemistry.

The axial hydrogen atoms at the ring junctions have a *trans-* relationship

11.3   Amino acids, peptides and proteins

Peptides and proteins are composed of amino acids linked by amide (or peptide) bonds. There are 20 common amino acids, which are naturally occurring, and these all contain an amine group at the α-carbon of a carboxylic acid. All of the 20 most common amino acids contain primary amine groups, except proline, which is a cyclic amino acid bearing a secondary amine.

R = CH(OH)Me, threonine
R = CH$_2$SH, cysteine
R = CH$_2$CH$_2$SMe, methionine
R = CH$_2$CO$_2$H, aspartic acid

R = H, glycine
R = Me, alanine
R = CHMe$_2$, valine
R = CH$_2$CHMe$_2$, leucine
R = CH(Me)Et, isoleucine
R = CH$_2$OH, serine

R = CH$_2$CH$_2$CO$_2$H, glutamic acid
R = CH$_2$CONH$_2$, asparagine
R = CH$_2$CH$_2$CONH$_2$, glutamine
R = (CH$_2$)$_4$NH$_2$, lysine
R = (CH$_2$)$_3$NHC(=NH)NH$_2$, arginine
R = CH$_2$Ph, phenylalanine

R = [tryptophan structure] tryptophan   R = H$_2$C—⟨benzene⟩—OH tyrosine

R = [histidine structure] histidine   [proline structure] proline

The primary amino acids differ in the nature of the alkyl side chain (R). All of the 20 common amino acids except glycine (R = H) are chiral, and only one enantiomer is produced in nature. Most amino acids found in nature have the L-configuration (because of their stereochemical similarity to L-glyceraldehyde). All of the 20 common amino acids have the S configuration, except for cysteine.

Fischer projections

[L-alanine structure]
L-alanine
S configuration

CO$_2$H
H$_2$N—H
CH$_3$

CHO
HO—H
CH$_2$OH
L-glyceraldehyde

Amino acids exist as *zwitterions*, as they contain both a positive and negative charge within the same molecule. They are *amphoteric*, as they can react with acid to gain a proton, or with base to lose a proton.

[structure at low pH]   [neutral zwitterion structure]   [structure at high pH]

low pH   *neutral zwitterion*   *high pH*

The pH at which the amino acid exists primarily as the neutral zwitterion is known as the *isoelectric point*.

The alkyl side chains (R) can be divided into neutral, acidic and basic side chains. Those with alkyl or aryl side chains are neutral, those with amine (or related) side chains are basic, while those with carboxylic acid side chains are acidic.

Amino acids join together to form the (amide or) *peptide bond* of peptides and proteins. Dipeptides and tripeptides are formed, for example, by the combination of two and three amino acids, respectively. The individual amino acids within the peptide are known as *residues*, and whereas polypeptides usually contain less than 50 residues, proteins often contain more than 50 residues.

Peptides are written with the N-terminus on the left and the C-terminus on the right. The repeating sequence of nitrogen, α-carbon and carbonyl groups is known as the *peptide backbone*. The amino acid sequence determines the structure of the peptide/protein. As the peptide/protein size increases, so does the number of possible amino acid combinations. For example, for a protein containing 300 residues, there are 20^{300} possible amino acid combinations (as there are 20 common amino acids).

The shape of peptides/proteins is crucial for their biological activity. Although the side chains of the residues are free to rotate, this is not possible for the peptide bonds. This is because the nitrogen lone pair (of the peptide bond) is conjugated with the carbonyl group. Therefore, the C−N bond has partial double-bond character, and this slows down the rate of rotation around the C−N bond, which helps make the peptides/proteins relatively rigid.

The sequence of amino acids in a protein is called the *primary structure*, while the localised spatial arrangement of amino acid segments is called the *secondary structure*. The secondary structure results from the rigidity of the amide bond and any other non-covalent interactions (e.g. hydrogen bonding) of the side chains. Secondary structures include the *α-helix* and the *β-pleated sheet*. The *tertiary structure* refers to the way in which the entire protein is folded into a 3-dimensional shape, and the *quaternary structure* refers to the way in which proteins come together to form aggregates.

Large proteins, which act as catalysts for biological reactions, are known as *enzymes*. The tertiary structure of enzymes usually produces 3-dimensional pockets known as the *active sites*. The size and shape of the active site is specific for only a certain type of substrate, which is selectively converted to the product by the enzyme. This is often compared with a key fitting a lock (*the lock and key model*). The catalytic activity of the enzyme is destroyed by *denaturation*, which is the breakdown of the tertiary structure (i.e. the protein unfolds). This can be caused by a change in temperature or pH.

11.4 Nucleic acids

The nucleic acids DNA (deoxyribonucleic acid) and RNA (ribonucleic acid) carry the cell's genetic information. Indeed, DNA contains all the information needed for the survival of the cell.

- Both DNA and RNA are composed of phosphoric acid, a sugar and several heterocyclic organic bases. DNA contains the sugar deoxyribose and RNA contains the sugar ribose. The bases adenine (A), guanine (G), cytosine (C) and thymine (T) are present in DNA, while adenine (A), guanine (G), cytosine (C) and uracil (U) are present in RNA.

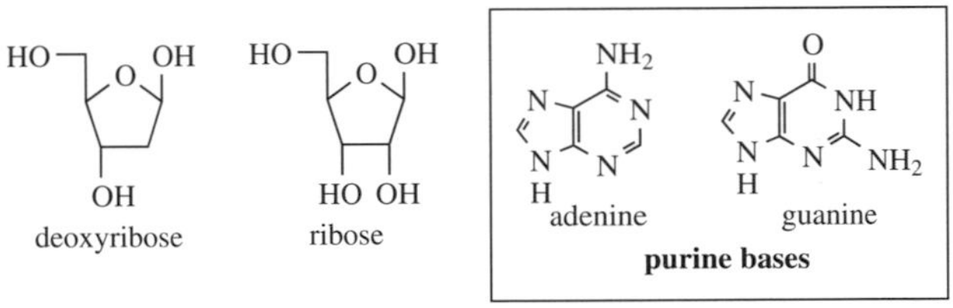

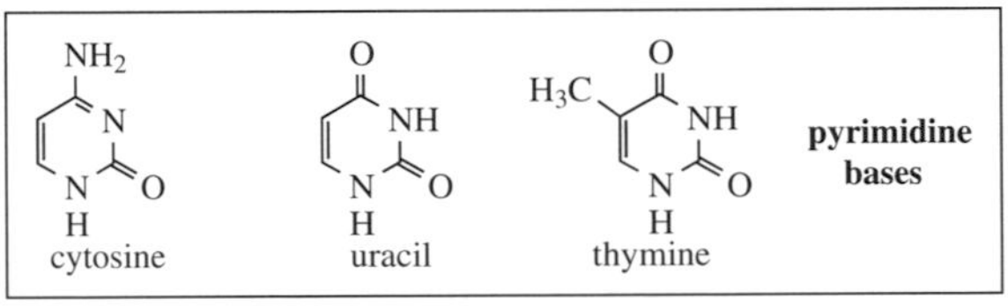

- Both DNA and RNA are polymers of *nucleotides* (phosphate–sugar–base), which are formed from *nucleosides* (sugar–base) and phosphoric acid. However, the polymer chain of DNA is much larger than that of RNA.

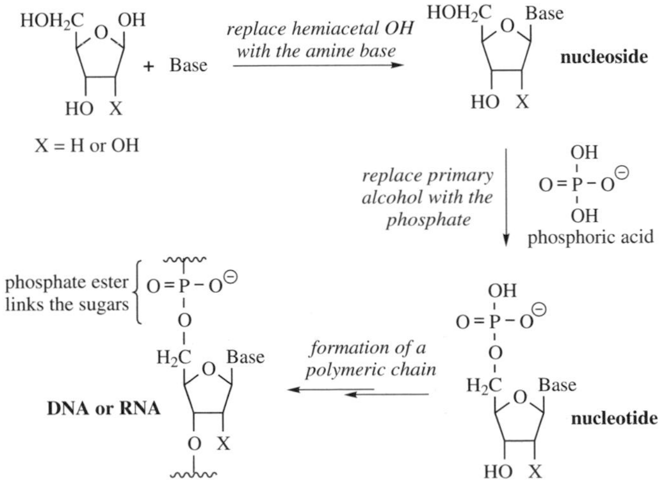

The structures of both DNA and RNA depend on the sequence of the nucleotides (i.e. the amine bases). Watson and Crick showed that DNA is a *double helix* composed of two strands with complementary bases, which hydrogen bond to one another. A and T form strong bonds to one another, as does C and G.

A ::::::::::::::: T
two hydrogen bonds

G ::::::::::::::: C
three hydrogen bonds

RNA is formed by the *transcription* of DNA. On cell division, the two chains of the helix unwind, and each strand is used as a template for the construction of an RNA molecule. The complementary bases pair up, and the completed RNA (which corresponds to only a section of the DNA) then unwinds from DNA and travels to the nucleus. Unlike DNA, RNA remains a single strand of nucleotides.

11.5   Synthetic polymers

A *polymer* is a large molecule made up of a repeating sequence of smaller units called *monomers*. Naturally occurring polymers include DNA and also cellulose, which is composed of repeating glucose units (see Sections 11.1 and 11.4). Synthetic polymers, which are made on a large scale in industry, have found a variety of important applications, e.g. adhesives, paints and plastics.

Polymers can be divided into *addition* (or chain-growth) polymers, formed on simple addition of monomers, or *condensation* (or step-growth) polymers, formed on the addition of monomers and elimination of a by-product such as water.

11.5.1   Addition polymers

Addition polymers can be formed by the reaction of an alkene with a radical, cation or anion initiator.

- *Radical polymerisation* is the most important method, and this often employs a peroxide (ROOR) initiator containing a weak oxygen–oxygen bond. On homolysis (using heat or light), the resulting alkoxyl radical (RO•) adds to the least hindered end of the alkene to form a carbon-centred radical. This radical then adds to another molecule of the alkene to build the polymer by a chain reaction. The polymerisation is terminated by, for example, the coupling of two radicals.

Initiation

$$RO \overset{\frown}{\frown} OR \xrightarrow{\text{heat or light}} RO^{\bullet} + {}^{\bullet}OR$$

Propagation

$$RO^{\bullet} \overset{\frown}{\frown} \underset{R}{\diagup} \longrightarrow RO \underset{R}{\diagup}^{\bullet} \overset{\frown}{\diagup} \underset{R}{\diagup} \xrightarrow{\substack{\text{further} \\ \text{alkene}}} RO\left[\underset{R}{\diagup}\underset{R}{\diagup}\right]_n^{\bullet}$$

Termination

$$RO\left[\underset{R}{\diagup}\underset{R}{\diagup}\right]_n^{\bullet} \overset{\frown}{\frown} {}^{\bullet}\left[\underset{R}{\diagup}\underset{R}{\diagup}\right]_n OR \longrightarrow RO\left[\underset{R}{\diagup}\right]_n\left[\underset{R}{\diagup}\right]_n OR$$

- *Cationic polymerisation* employs a strong protic or Lewis acid initiator. A proton, for example, adds to an electron-rich alkene to form the most stable carbocation. The carbocation then adds to another electron-rich alkene to build the polymer chain. The polymerisation is terminated by, for example, deprotonation.

Propagation

$$H^{\oplus} \overset{\frown}{\diagup} \underset{R}{\diagup} \longrightarrow H \underset{R}{\diagup}^{\oplus} \overset{\frown}{\diagup} \underset{R}{\diagup} \xrightarrow{\substack{\text{further} \\ \text{alkene}}} H\left[\underset{R}{\diagup}\underset{R}{\diagup}\right]_n^{\oplus}$$

R = electron-donating group

Termination

$$H\left[\underset{R}{\diagup}\underset{R}{\diagup}\right]_n^{\oplus} \xrightarrow{-H^{\oplus}} H\left[\underset{R}{\diagup}\right]_n \diagup \underset{R}{\diagdown}$$

- *Anionic polymerisation* employs nucleophiles such as alkyllithiums, alkoxides or hydroxide as the initiator. Hydroxide, for example, adds to an electron-deficient alkene to form the most stable carbanion (in a Michael-type reaction). The carbanion then adds to another electron-deficient alkene to build the polymer chain. The polymerisation is terminated by, for example, protonation.

Propagation

$$HO^{\ominus} \overset{\frown}{\diagup} \underset{R}{\diagup} \longrightarrow HO \underset{R}{\diagup}^{\ominus} \overset{\frown}{\diagup} \underset{R}{\diagup} \xrightarrow{\substack{\text{further} \\ \text{alkene}}} HO\left[\underset{R}{\diagup}\underset{R}{\diagup}\right]_n^{\ominus}$$

R = electron-withdrawing group

Termination

$$HO\left[\underset{R}{\diagup}\underset{R}{\diagup}\right]_n^{\ominus} \xrightarrow{H^{\oplus}} HO\left[\underset{R}{\diagup}\underset{R}{\diagup}\right]_n H$$

Radical polymerisation of alkenes often leads to chain branching and the formation of non-linear polymers. In addition, the stereogenic centres on the backbone of polymers are usually formed randomly to give *atactic* polymers.

atactic (random)

isotactic (same side)

syndiotactic (alternating)

The formation of linear *isotactic* or *syndiotactic* polymers can be achieved by metal-catalysed polymerisation. This employs *Ziegler–Natta catalysts*, made from triethylaluminium (Et_3Al) and titanium tetrachloride ($TiCl_4$), which react with alkenes by a complex mechanism. The polymerisation of ethylene ($CH_2{=}CH_2$) leads to the formation of (linear) high-density polyethylene, which is of greater strength than the (branched) low-density polyethylene produced on radical polymerisation.

Copolymers can be formed when two or more monomers polymerise to give a single polymer. These polymers often exhibit properties different from that of *homopolymers* (produced from only one alkene).

11.5.2   Condensation polymers

Condensation polymers are formed from two monomers containing two functional groups. The formation of a bond usually leads to the elimination of a simple by-product (such as water), and this occurs in discrete steps (i.e. not *via* a chain reaction). The polymer is usually composed of an alternating sequence of the two monomers.

- *Polyamides* (or nylons) are produced by heating diacids with diamines. The amine acts as the nucleophile and reacts with the carboxylic acid in a nucleophilic acyl substitution reaction (see Section 9.3).

- *Polyesters* are produced by heating diacids (or diesters) with diols.

- *Polyurethanes* are produced by the reaction of diols with diisocyanates. Nucleophilic addition of an alcohol to an isocyanate produces the urethane functional group. Although no (small molecule) by-products are produced, the urethane bonds are formed in discrete steps.

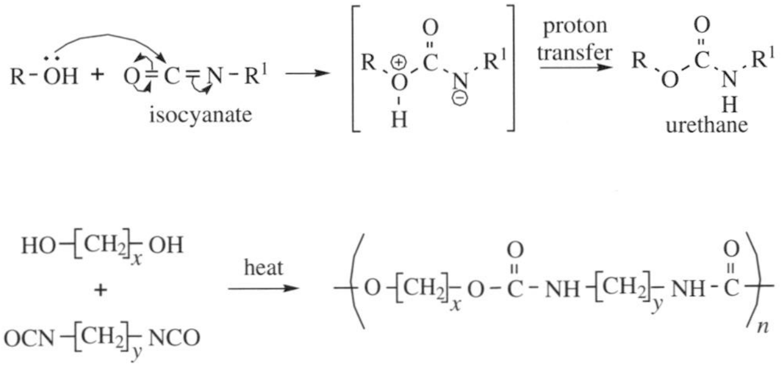

Problems

(1) Draw structures of the two possible dipeptides, which can be formed by joining (*S*)-alanine to glycine.

(2) The following questions relate to the disaccharide cellobiose (**A**).

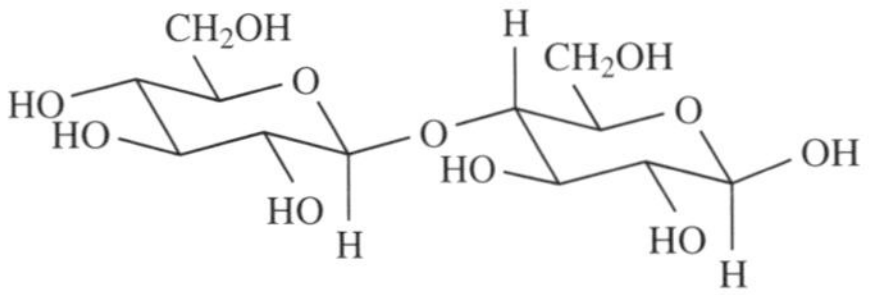

A

(a) Classify the glycoside linkage as either α or β.

(b) Name the monosaccharide(s) formed when **A** is hydrolysed in aqueous acid.

(c) Is the monosaccharide(s) formed on the hydrolysis of **A** an aldohexose or a ketohexose?

(3) Are the OH and Me groups in deoxycholic acid (**B**) (a bile acid) axial or equatorial?

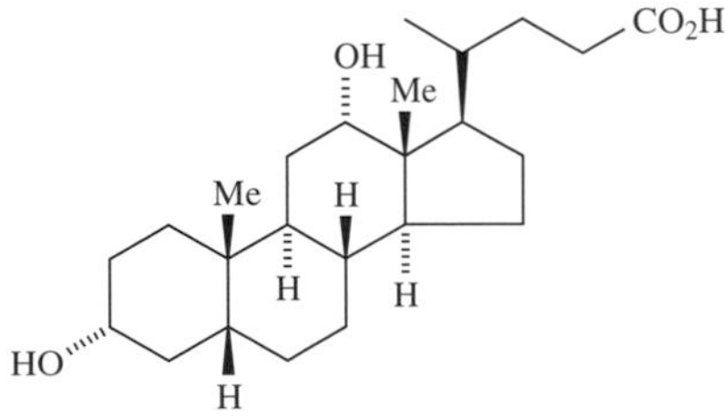

B

(4) How can tautomerism explain why nucleic acids, such as uracil (**C**) and guanine (**D**), are aromatic even though this is not indicated in the structures shown below?

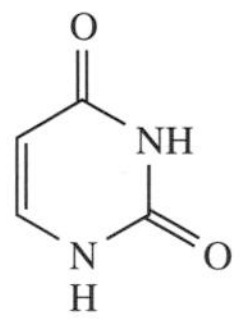

C 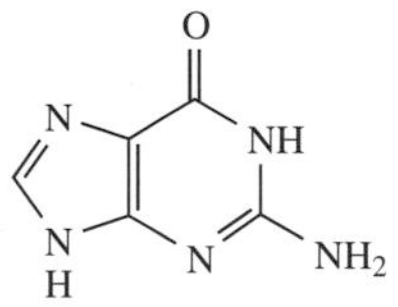   D

(5) Why is radical polymerisation of vinyl chloride to give poly(vinyl chloride) described as showing a marked preference for head-to-tail addition?

(6) What polymers would you expect to be formed from the following reactions?

(a)

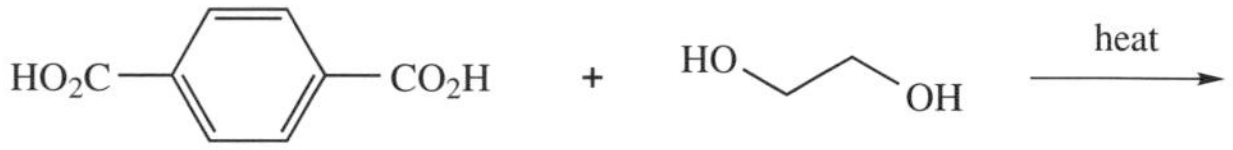

(b)

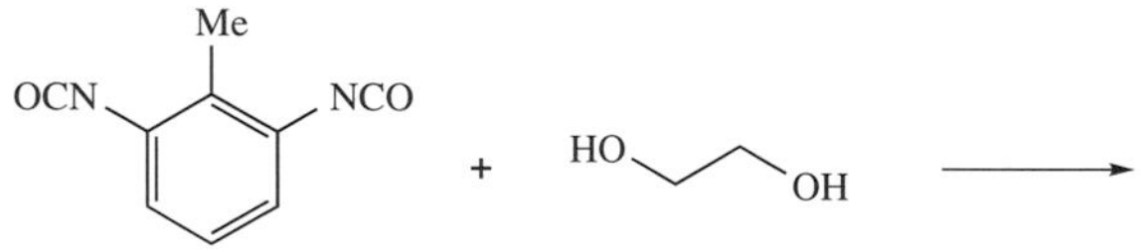

OUTLINE ANSWERS

1. Structure and Bonding

(1) (a) +I   (b) −I, +M (c) −I, +M (d) −I, +M
     (e) −I, +M (f) −I, −M (g) −I, −M (h) −I, −M

(2) (a)

A

B

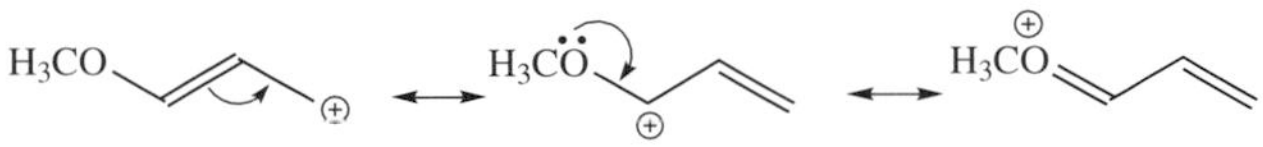

C

(b) Expect **C** to be the more stable because of the +M effect of the OMe substituent.

(3) (a) The carbocation $CH_3OCH_2^+$ is stabilised by the +M effect of the OCH_3 substituent.

(b) The electron-withdrawing NO_2 (−I, −M) group can stabilise the phenoxide anion by delocalisation of the negative charge. The negative charge can be spread on to the NO_2 group at the 4-position.

(c) On deprotonation of CH_3COCH_3, the anion (known as an enolate anion) can be stabilised by delocalisation of the negative charge onto the C=O group. As a consequence, CH_3COCH_3 is more acidic than CH_3CH_3.

(d) This can be explained by resonance. Drawing a second resonance structure for CH_2=CH−CN shows that the C−C bond in CH_2=CH−CN has a partial double bond character (see below). This is not possible for CH_3−CN.

(e) For CH_2=CH−CH_2^+, the carbon atom bearing the positive charge has six valence electrons. It can accept two further electrons to generate an alternative resonance structure.
For CH_2=CH−NMe_3^+, the nitrogen atom bearing the positive charge has eight valence electrons. It cannot expand its valence electrons to ten, and hence an alternative resonance structure cannot be drawn.

(4)   On deprotonation of cyclopentadiene, an anion with six π-electrons is formed. This anion is stabilised by aromaticity, hence the low pK_a-value of cyclopentadiene.
      On deprotonation of cycloheptatriene, an anion with eight π-electrons is formed. This anion is not stabilised by aromaticity, hence the high pK_a-value of cycloheptatriene.

(5)   The most acidic hydrogen atom in each compound is shown in **bold**. (Approximate pK_a-values for these hydrogen atoms are given in brackets.)
      (a)   4-**HO**$C_6H_4CH_3$ (pK_a ~10).
      (b)   4-HO$C_6H_4CO_2$**H** (pK_a ~5).
      (c) H_2C=CHCH$_2$CH$_2$C≡C**H** (pK_a ~25).
      (d)   **H**OCH$_2$CH$_2$CH$_2$C≡CH (pK_a ~16).

(6)   (a)   guanidine > 1-aminopropane > aniline > ethanamide
      On protonation of guanidine, the $(H_2N)_2C=NH^+$ cation is stabilised by extensive delocalisation of the charge by the electron-donating NH_2 (+M) groups (see below).

      The electron-donating propyl group (+I) in 1-aminopropane will help increase the basicity.
      For aniline, the lone pair on nitrogen can be delocalised onto the benzene ring, which will help decrease the basicity (i.e. the benzene ring helps 'tie up' the lone pair, making it less basic). On protonation of aniline, the positive charge cannot be stabilised by delocalisation.
      Ethanamide is the weakest base because of the delocalisation of the nitrogen lone pair with the electron-withdrawing carbonyl group. It should be noted that if ethanamide was protonated on nitrogen, the positive charge could not be stabilised by delocalisation. Protonation therefore occurs on oxygen, as the charge can be delocalised onto the nitrogen atom (see below).

      (b)   4-methoxyaniline > 4-methylaniline > aniline > 4-nitroaniline
      For 4-methoxyaniline, the electron-donating OMe group (+M) increases the basicity of the amine more than the 4-methyl group (+I) in 4-methylaniline. Both amines are more basic than aniline, as they contain electron-donating groups at the 4-position of the ring. In contrast, the strongly electron-withdrawing nitro group (−I, −M) reduces the basicity of the amine, and hence 4-nitroaniline is much less basic than aniline.

2. Functional Groups, Nomenclature and Drawing Organic Compounds

(1)

(a)   (b)   (c)

(d)   (e)   (f)

(this could be E or Z;   (this could be E or Z;
see Section 3.3.1)   see Section 3.3.1)

(g)   (h)

(i)   (j)

(2) (a) 3-Hydroxy-2-methylpentanal.
    (b) Ethyl 4-aminobenzoate.
    (c) 1-Ethynylcyclohexanol.
    (d) 4-Cyanobutanoic acid.
    (e) 3-Bromo-1-methylcyclopentene.
    (f) 5-Methyl-2-(1-methylethyl)phenol or 2-isopropyl-5-methyl-
         phenol (isopropyl = 1-methylethyl = $CHMe_2$).
    (g) N-1,1-Dimethylethylpropanamide   or N-$tert$-butylpropan-
         amide ($tert$-butyl = 1,1-dimethylethyl = CMe_3).
    (h) 4-Chloro-6-methyl-3-heptanone.
    (i) Benzylcyclohexymethylamine
         (benzyl = phenylmethyl = $PhCH_2$).
    (j) 2,5-Dimethylhex-4-en-3-one.

3. Stereochemistry

(1) (a) $-I$ (highest), $-CH_3$ (lowest).
  (b) $-CO_2H$.
  (c) $A = (E)$, $B = (E)$, $C = (Z)$.

(2) $D = (S)$, $E = (R)$, $F = (R)$.

(3) (a) It is optically inactive but contains stereogenic centres. As **G** has a plane of symmetry, the optical activity of one half of the molecule cancels out the other half.

  (b)

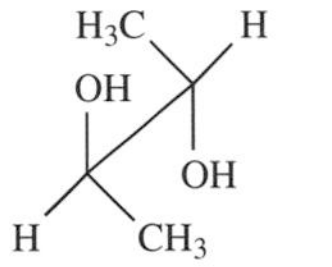

Sawhorse projection 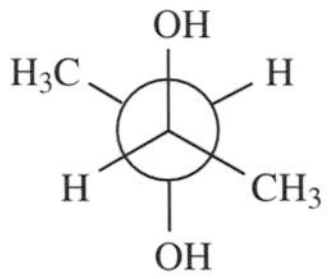   Newman projection

  (c)

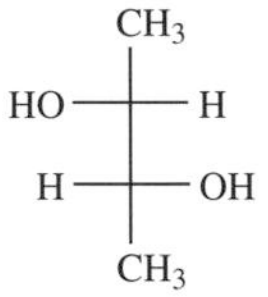

(4)

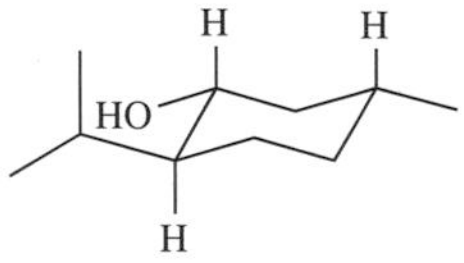

All three substituents are equatorial

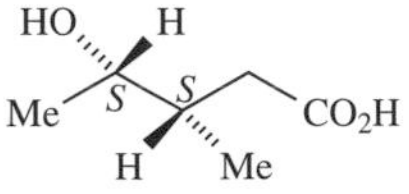

(5)

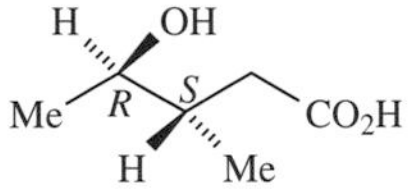

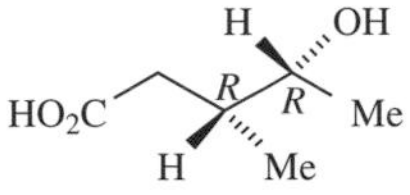

I and J are diastereoisomers
I and K are diastereoisomers

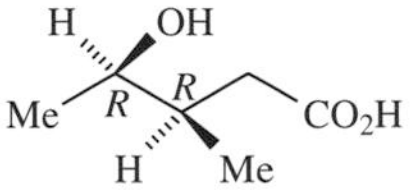

I and **L** are diastereoisomers
J and **K** are enantiomers
J and **L** are identical
K and **L** are enantiomers

(6)

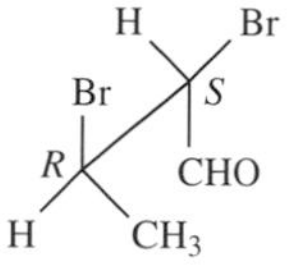

M

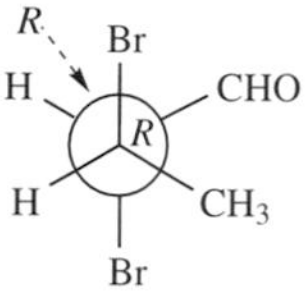

N

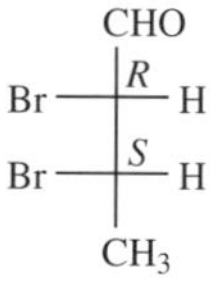

O

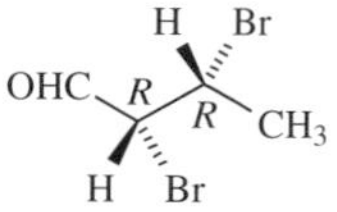

P

M and **N** are diastereoisomers
M and **O** are enantiomers
M and **P** are diastereoisomers
N and **O** are diastereoisomers
N and **P** are identical
O and **P** are diastereoisomers

4. Reactivity and Mechanism

(1)   Oxidation levels are shown in italic (see below).

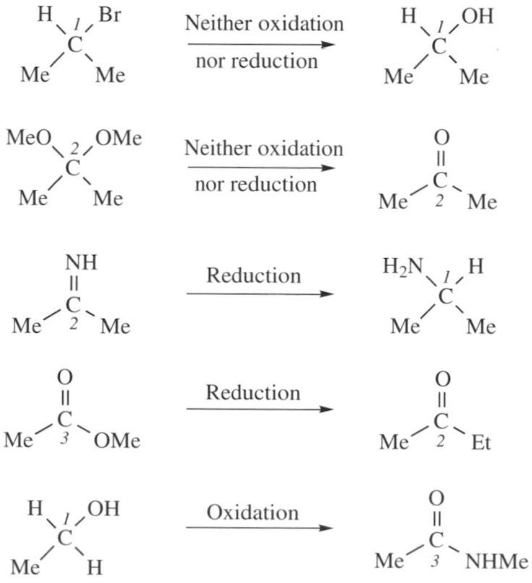

(2)

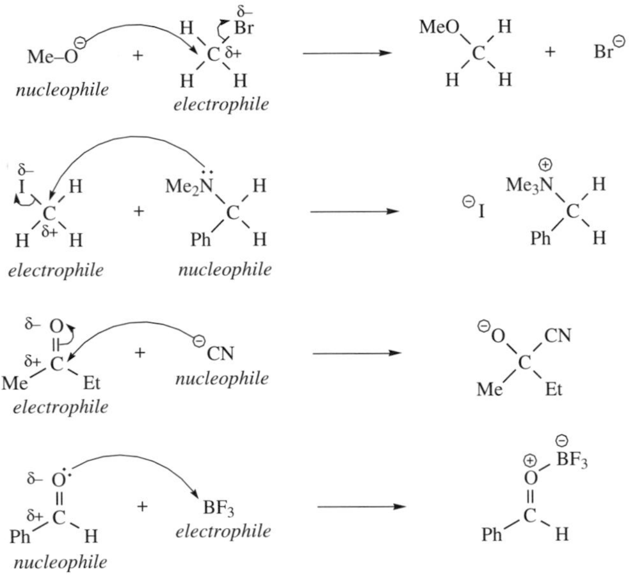

(3) (a) (radical) substitution.
    (b) addition.
    (c) two eliminations.
    (d) (Beckmann) rearrangement.

(4) (a) stereoselective (forming the *E*-isomer from the elimination).
    (b) regioselective (elimination to give the less substituted alkene).
    (c) chemoselective (reduction of the carboxylic acid).
    (d) stereoselective (reduction of the ketone to selectively form one diastereoisomer).

5. Alkyl Halides

(1) (a)

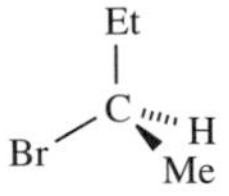

   (b) Ethanol can act as a nucleophile and react with the secondary bromide to form an ethyl ether (see below) together with HI. As ethanol is a polar protic solvent, the reaction is likely to involve an S_N1 reaction to give a racemic product.

$$\underset{\text{EtO}}{\overset{\overset{\displaystyle \text{Et}}{\underset{\displaystyle |}{}}}{\text{C}}}\overset{\text{H}}{\underset{\text{Me}}{}}$$

(c) Hydroxide is only a moderate nucleophile but a good base. Elimination can therefore compete with substitution, leading to the formation of alkene products.

(d) For an S_N1 reaction, the rate-determining step is the formation of a carbocation. Whereas **A** produces a secondary alkyl carbocation, S_N1 reaction of (1-iodoethyl)benzene produces a benzylic carbocation, $PhCH^+CH_3$. This benzylic carbocation is stabilised by resonance and should be formed more readily than $EtCH^+CH_3$.

(2) (a)

Br H

(b)

S_N1

Br H $-Br^{\ominus}$ $\oplus$ $H_2\ddot{O}$ OH

then $-H^{\oplus}$

racemic

S_N2

$H_2\ddot{O}$ + H Me $-Br^{\ominus}$ OH

Br then $-H^{\oplus}$ H Me

(R)-2-bromopentane   (S)-pentan-2-ol

(c) Adding a less polar non-protic solvent such as propanone could promote an S_N2 reaction.

(d) This is an example of nucleophilic catalysis, which exploits the fact that I^- is a good nucleophile and a good leaving group. Reaction of **B** with the iodide ion in an S_N2 reaction produces 2-iodopentane, which reacts more rapidly with water than does 2-bromopentane (**B**).

(e) Sodium *tert*-butoxide is a poor nucleophile but a strong base, and hence elimination rather than substitution is favoured. As Me_3C-O^- is a bulky base, pent-1-ene is likely to be the major product (from a Hofmann elimination).

(f) On protonation, the OH group of **C** is converted into a better leaving group, namely water. Loss of water produces a carbocation, which can rapidly react with ethanol in an S_N1 reaction.

(3) (a) $PhCH_2OMe + HBr$

(b) $PhCH_2OCOCH_3 + HBr$

(4)   The mechanism of the reaction changes from S_Ni in the absence of pyridine to S_N2 in the presence of pyridine (which acts as a base).

(5)

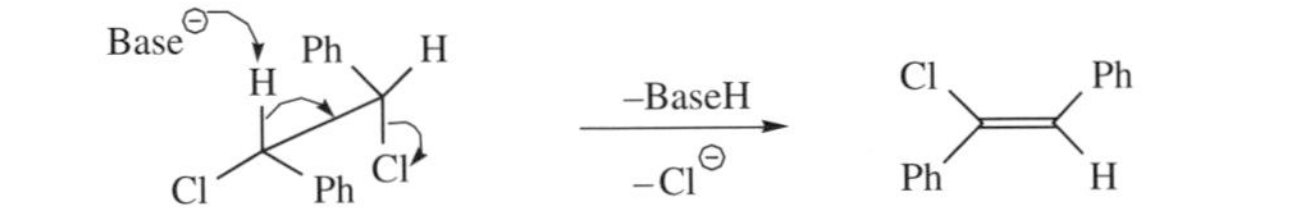

(1S,2S)-1,2-dichloro-1,2-diphenylethane   (Z)-1-chloro-1,2-diphenylethane

(6)   Ethanol is a poor nucleophile and a weak base, and hence S_N2 and E2 reactions are unlikely. As it is a polar protic solvent, E1 and S_N1 reactions are favoured and hence an initial carbocation can be formed on the loss of Br$^-$ (see below). This secondary carbocation can undergo rearrangement to form a more stable tertiary carbocation, which can react with ethanol to give **E** and **F**.

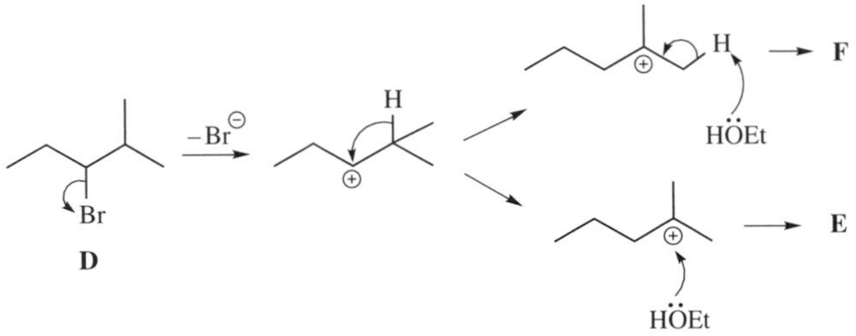

6. Alkenes and Alkynes

(1)   (a)   This involves a radical mechanism in which the bromine radical adds to the least hindered end of the alkene to form the more stable carbon-centred radical (i.e. a secondary rather than a primary radical).

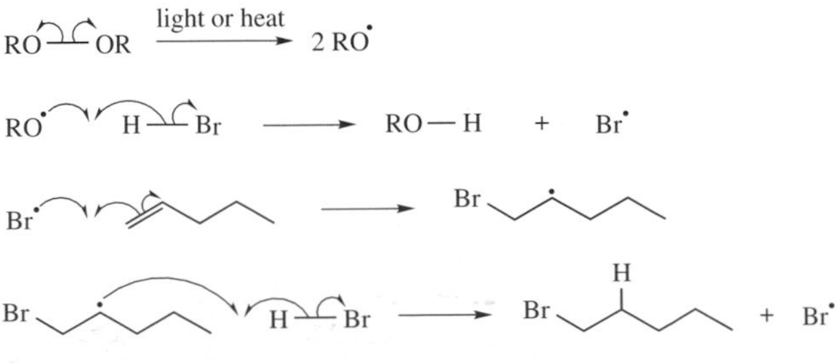

      (b)   This involves an ionic mechanism in which protonation of the alkene produces the more stable carbocation intermediate (i.e. a secondary rather than a primary carbocation).

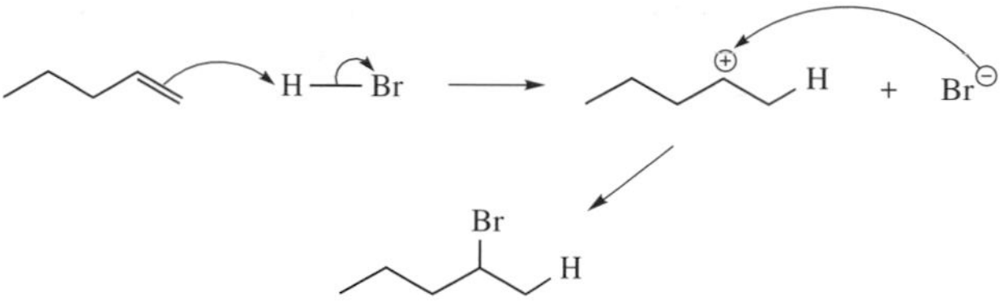

(2) The reaction involves an intermediate bromonium ion, which can be ring-opened by attack of Br⁻ at either carbon atom of the 3-membered ring (as these are equivalent). The bridging bromine atom prevents rotation about the central C–C bond, and hence the Z-relationship of the ethyl groups in the alkene is conserved in the bromonium ion. Overall, there is *anti-* addition of Br₂ to give a racemate.

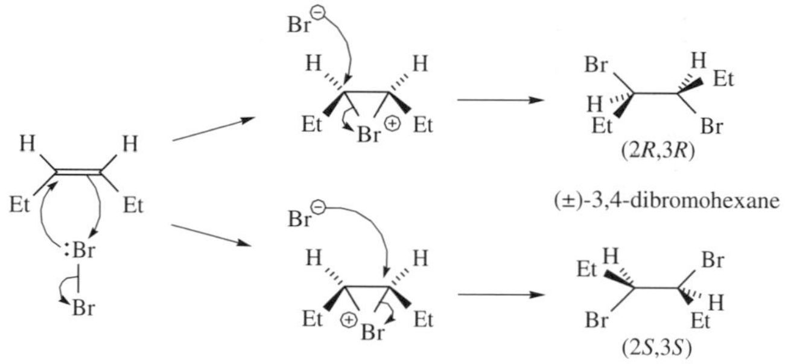

(±)-3,4-dibromohexane

(3) (a) RCO₃H, such as *meta*-chloroperoxybenzoic acid (*m*-CPBA).
   (b) RCO₃H and then ring-opening of the epoxide with HO⁻/H₂O or H⁺/H₂O to give stereoselective *anti-* dihydroxylation.
   (c) H₂/Pd/C.
   (d) OsO₄ then H₂O/NaHSO₃ (to give stereoselective *syn-* dihydroxylation).
   (e) BH₃ (regioselective hydroboration) then H₂O₂/HO⁻ (H and OH are added in a *syn-* manner).
   (f) O₃ (ozonolysis) then H₂O₂ (oxidative work-up).

(4) **A** Hg(OAc)₂/H₂O/H⁺ (hydration).
   B (i) NaNH₂ then MeBr; (ii) Na/NH₃.
   C (i) NaNH₂ then PhCH₂Br; (ii) H₂/Lindlar catalyst.

(5)

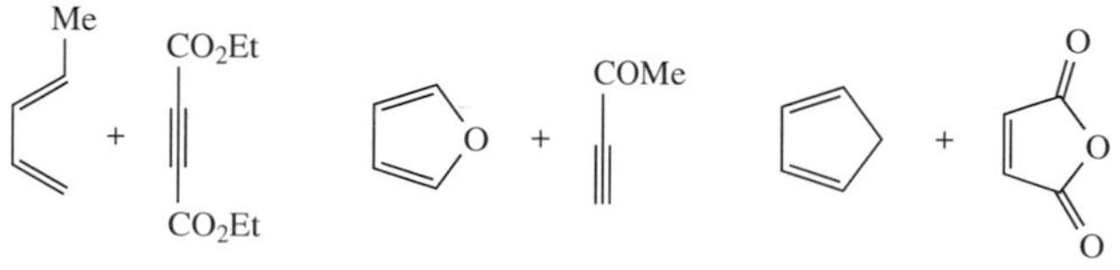

D   **E**   **F**

For compound **F**, the so-called *endo-* product is formed. For the *endo-* product, the newly formed alkene and carbonyl group(s) are on the same side of the product. *Endo-* products are often formed more rapidly than *exo-* products in Diels–Alder reactions.

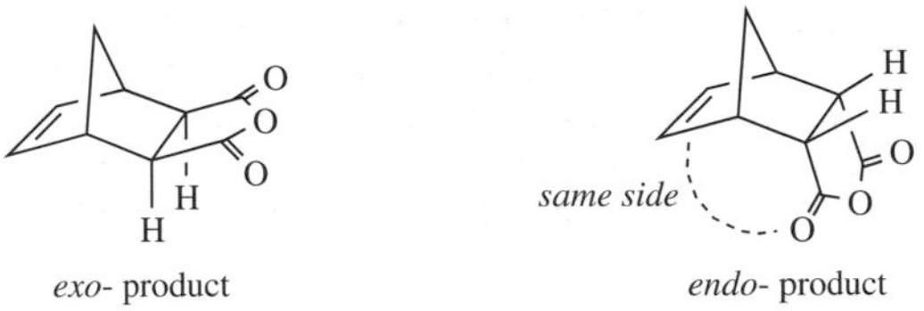

exo- product   *endo-* product

(6) On protonation of one of the double bonds, an allylic cation is formed and Cl⁻ can react with this in two possible ways (see below). 3-Chlorobut-1-ene is the expected Markovnikov addition product resulting from 1,2-addition. 1-Chlorobut-2-ene is derived from the trapping of the cation at the terminal carbon in a 1,4-addition.

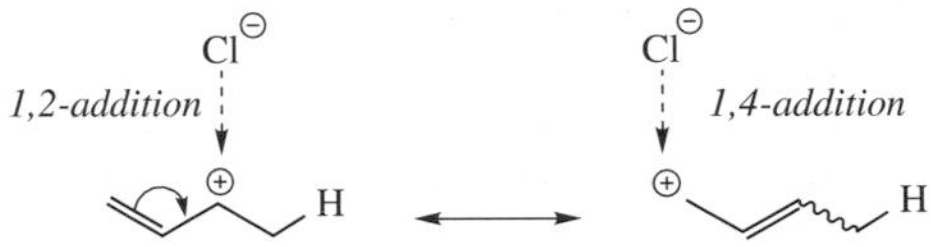

7. Benzenes

(1)

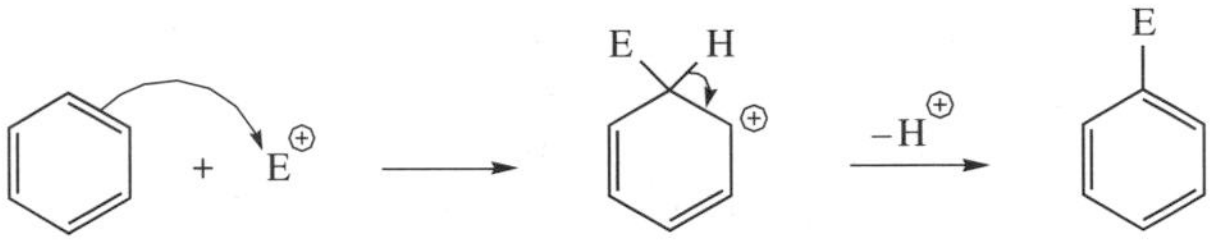

(2)

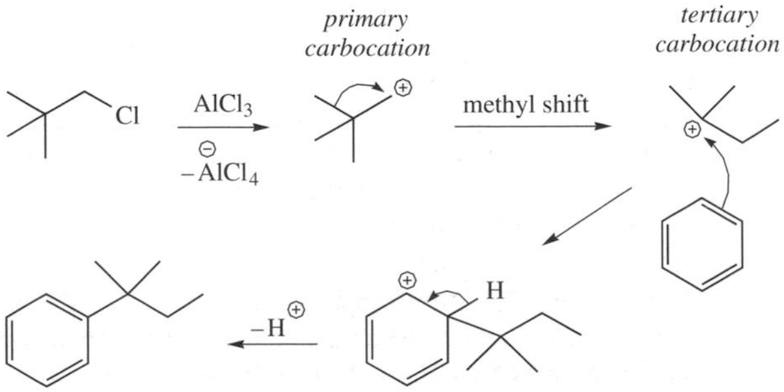

(3) (a)   **A** (cyclobuta-1,3-diene) is anti-aromatic (four π-electrons).
          B (anthracene) is aromatic (14 π-electrons).
          C (indole) is aromatic (10 π-electrons).
          D is aromatic (10 π-electrons).

(b)   This can be investigated by the reaction of **C** with an electrophile. Aromatic compounds undergo electrophilic substitution rather than addition reactions. On reaction with an electrophile such as bromine, indole undergoes an electrophilic substitution reaction to form 3-bromoindole (see below). As for pyrrole and furan, no Lewis acid is required for bromination.

(4) (a)   CHCl$_3$ and three equivalents of benzene (Friedel–Crafts alkylations).

(b)   (i) CH$_3$CH$_2$CH$_2$Cl/AlCl$_3$ or CH$_3$CH$_2$COCl/AlCl$_3$ followed by Zn/Hg/H$^+$; (ii) NBS/peroxide (bromination at the benzylic position).

(c)   (i) Br$_2$/FeBr$_3$; (ii) CH$_3$Cl/AlCl$_3$ (separate 1-bromo-4-methylbenzene from 1-bromo-2-methylbenzene); (iii) KMnO$_4$ (oxidation of the methyl group).

(d)   (i) Cl$_2$/FeCl$_3$; (ii) two equivalents of HNO$_3$/H$_2$SO$_4$ (to make 1-chloro-2,4-dinitrobenzene); (iii) H$_2$N–NH$_2$ (nucleophilic aromatic substitution).

(5) (a)   The –NH$_2$ group is an activating group and directs the electrophile to the 2-, 4- and 6-positions of the ring. In acid conditions, phenylamine is protonated and the –NH$_3^+$ group is deactivating and directs the electrophile to the 3- and 5-positions of the ring.

(b)   The alkene (cyclohexene) is not aromatic and hence is a more reactive nucleophile than benzene. In phenol, the –OH group (+M) is a strongly activating substituent, and hence the ring is more reactive to electrophiles than benzene.

(c)   As bromination of phenylamine can occur at the 2- as well as the 4-position, a mixture of isomers is formed. Conversion of the amine group to a larger amide group (by reaction with an acid chloride) can block the 2-position and increase the proportion of substitution at the 4-position. In the final step, the amide is converted back to the amine (HO$^-$/H$_2$O).

(6) (a)   (i) NaNO$_2$/HCl; (ii) KI.

(b)   (i) Br$_2$ (separate 2-bromophenylamine from the desired 4-bromophenylamine; the proportion of 4-bromophenylamine can be increased by the introduction of an amide-blocking

group; see Question (5c)); (ii) NaNO$_2$/HCl; (iii) CuCl (Sandmeyer reaction).
(c)   (i) Br$_2$ (excess) to make 2,4,6-tribromoaniline; (ii) NaNO$_2$/HCl;
(iii) H$_3$PO$_2$.

8. Carbonyl Compounds: Aldehydes and Ketones

(1)   (a)

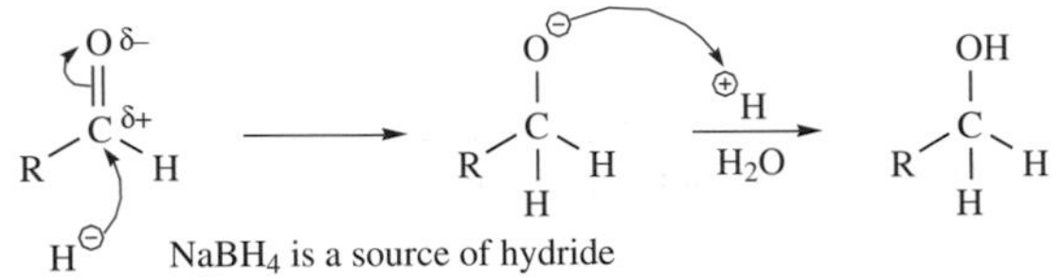

NaBH$_4$ is a source of hydride

(b)   Use CrO$_3$/H$^+$ (Jones oxidation) and distil the aldehyde as soon
as it is formed. Alternatively, use a milder oxidising agent
such as pyridinium chlorochromate (PCC) to prevent further
oxidation of the aldehyde, which would give a carboxylic acid.

(2)   (a)

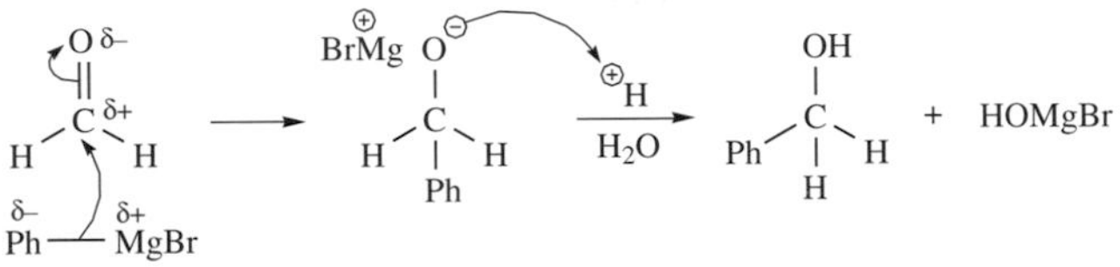

(b)   Ph−Br + Mg in anhydrous ethereal solvents (e.g. tetrahydro-
furan). Ideally, use an inert atmosphere, as Grignard reagents
can react with oxygen to form hydroperoxides (ROOH) and
alcohols.

(3)   A = CHI$_3$ (iodoform)   B = PhCO$_2$H (benzoic acid)

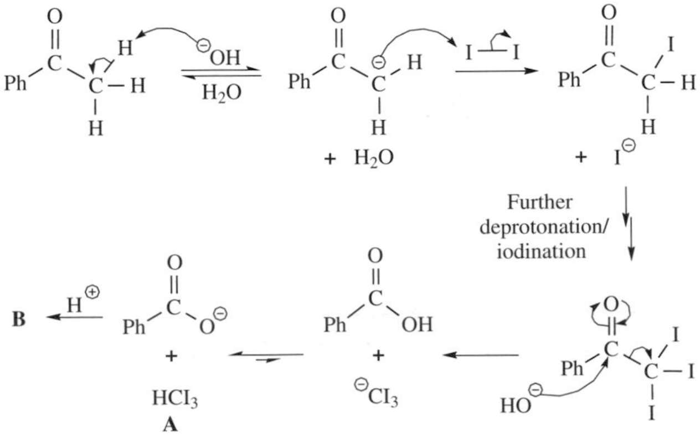

(4) (a)

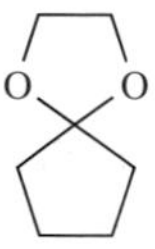

C

(b) The acid protonates the ketone, making it a better electrophile.
     This makes it more susceptible to nucleophilic attack by an
     alcohol.

(c)

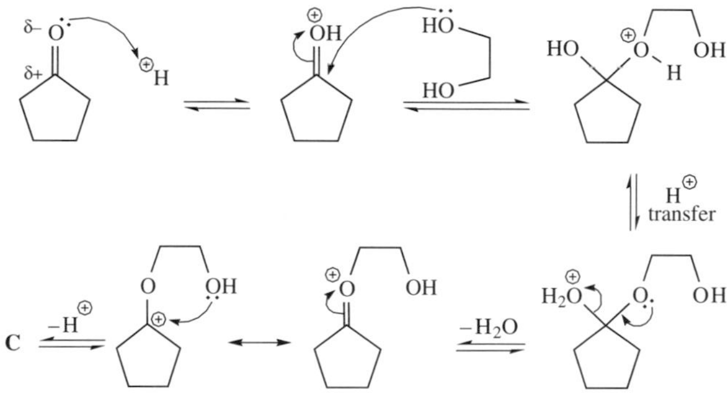

(5) (a)

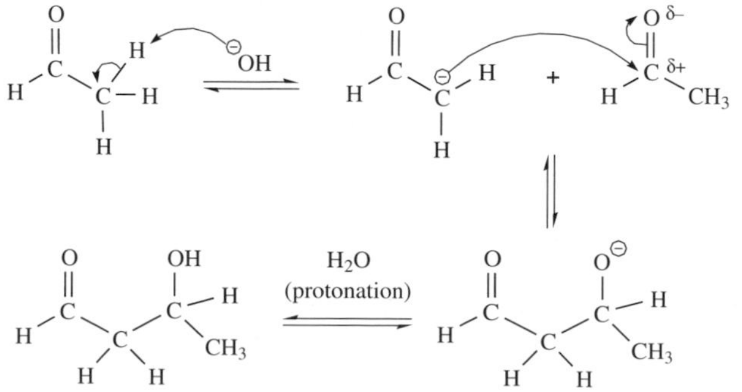

D

(b)

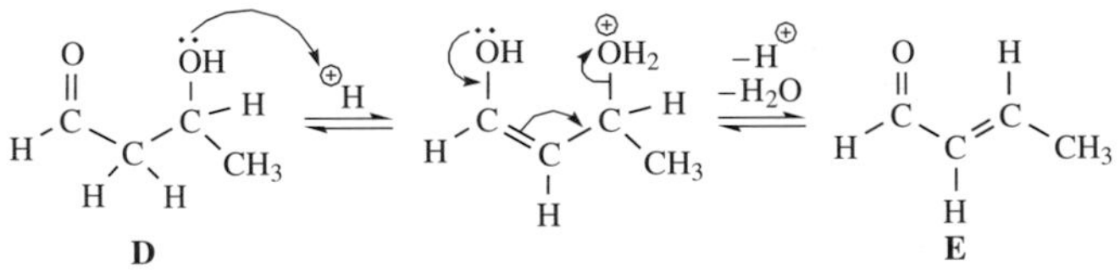

D   E

(6)

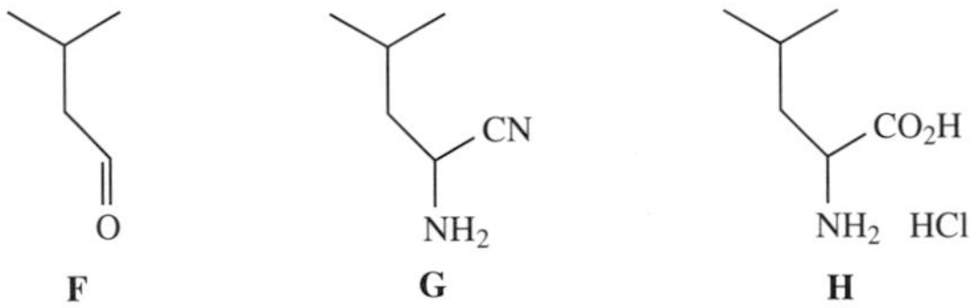

F   G   H

9. Carbonyl Compounds: Carboxylic Acids and Derivatives

(1) (a) NaBH$_4$ followed by H$^+$/H$_2$O.

The positive mesomeric effect (+M) of the OMe group ensures that the carbonyl carbon atom of the ester is less electrophilic than that of the ketone. Therefore, the ketone is more readily attacked by nucleophiles, and mild hydride reducing agents such as NaBH$_4$ only react at the ketone.

(b)

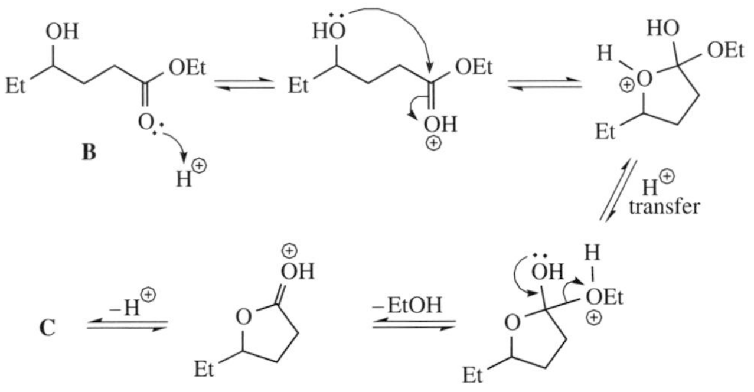

B

C

(c)

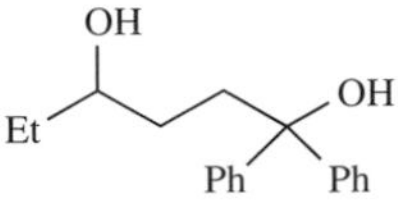

(2) (a) H$^+$/H$_2$O.
     (b) LiAlH$_4$ then H$_2$O.

(c) CrO_3/H^+ (to form $PhCO_2H$) then $EtOH/H^+$ (esterification).

(d) $EtOH/H^+$ (esterification).

(e) H_2O.

(f) $H^+/H_2O/heat$.

(g) $SOCl_2$ or PCl_3.

(h) $HNEt_2$.

(3) (a) $PhCOCH_2COPh$.

(b) $PhCOCH_2CHO$.

(c) $(CH_3)_3CCOCH_2CO_2CH_2CH_3$.

(d) $PhCOCH_2COCH_3$.

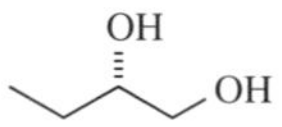

(4)

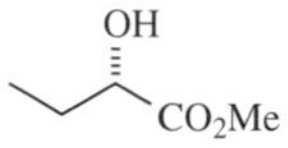

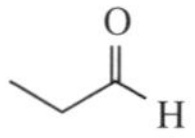

(5)

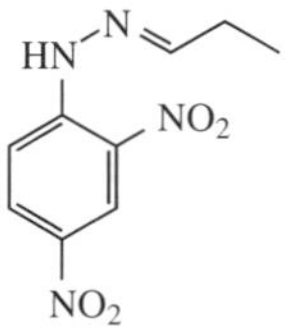

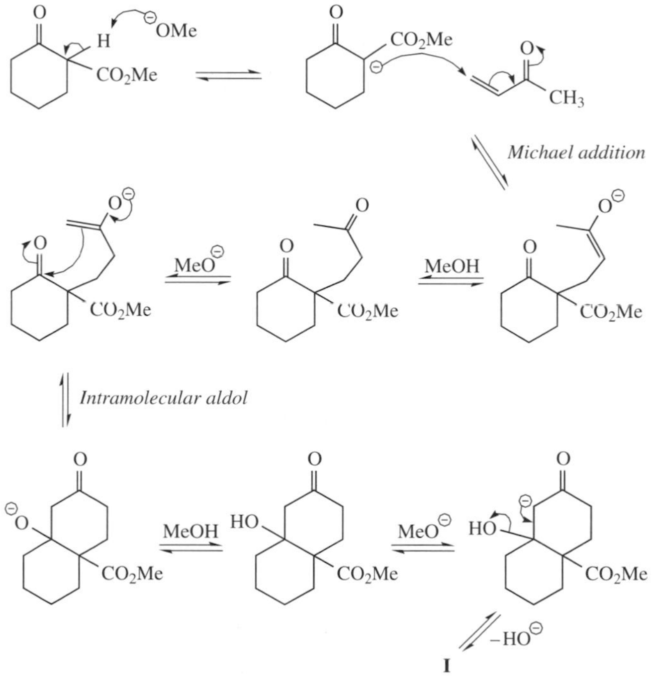

This sequence of reactions (i.e. Michael addition followed by intramolecular aldol condensation) to form a 6-membered ring is known as the *Robinson annelation*. Annelation refers to the formation of a ring.

10. Spectroscopy

(1)   (a)   Use high-resolution EI or CI mass spectrometry.

     (b)   From the characteristic 1:1 ratio of (M) and (M+2) peaks in the mass spectrum.

     (c)   A singlet at ~2.6 ppm due to the three equivalent protons of $-COCH_3$ would be observed in the ^1H NMR spectrum.

            In the ^{13}C NMR spectrum, a characteristic ketone peak would be observed around 200 ppm (the methyl carbon would give a peak at ~25 ppm).

            Fragmentation to form $4\text{-}BrC_4H_6C{\equiv}O^+$ could occur in the mass spectrum. This would help confirm the presence of an aromatic ketone, and the molecular ion peak could be used to establish that **A** is a methyl ketone.

     (d)   Two sets of distorted doublets (with the same *ortho* coupling constant of ~8 Hz) will be observed in the aromatic region of the ^1H NMR spectrum. This is because there is a plane of symmetry in the molecule, and hence two sets of protons are chemically equivalent. These are shown as H_A and H_M below. The $-COCH_3$ (−M, −I) group is more electron withdrawing than the −Br group (+M, −I) and hence the H_M protons have a higher chemical shift. This is known as an AM spectrum because the difference in chemical shift ($\Delta\delta$) of the two doublets is greater than the size of the coupling constant (J). For an AX spectrum, $\Delta\delta$ is much greater than J, and for an AB spectrum, J is greater than $\Delta\delta$. Notice that as the value of $\Delta\delta$ approaches the value of J, the inner lines increase in height to produce a 'two leaning doublet'.

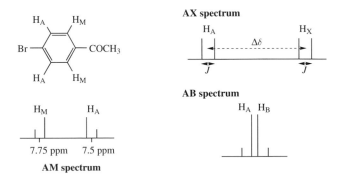

For the 1,2- or 1,3-disubstituted isomer of **A**, there will be four (not two) separate peaks in the aromatic region of the ^1H NMR spectra, because none of the aromatic protons are equivalent.

In the ^{13}C NMR spectrum for **A**, there will be four peaks in the aromatic region (two quaternary carbon peaks and two CH peaks). For a 1,2- or 1,3-disubstituted isomer of **A**, there will be six peaks in the aromatic region (two quaternary carbon peaks and four CH peaks), because none of the carbon atoms are equivalent.

(2)

O
Me

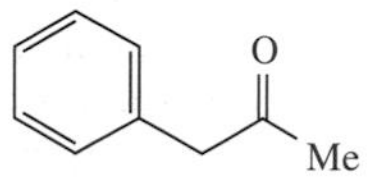

B

(3) **C** Approximate chemical shift values for **C**. Splitting patterns and coupling constants are given in brackets.

^{1}H NMR
4.1 (q, 7 Hz)
O
1.8 (d, 7 Hz)
4.5 (q, 7 Hz)
H
Cl
O
1.3 (t, 7 Hz)
^{13}C NMR
O
20
55
165
60
O
15
Cl

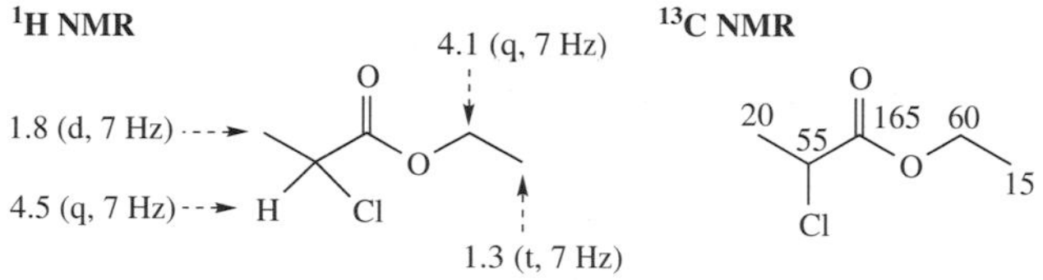

D Approximate chemical shift values for **D**. Splitting patterns and coupling constants are given in brackets.

^{1}H NMR
1.8 (t, 7 Hz)
1.0 (t, 7 Hz)
O
2.1 (s)
CHO
1.3 (sex, 7 Hz)
1.5 (s)
10.0 (s)
^{13}C NMR
O
210
75
30
20
15
20
10
CHO
205

(4)

O
H
OH
OMe

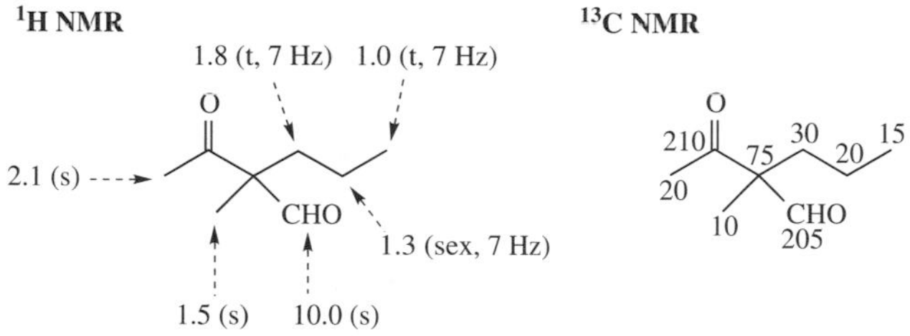

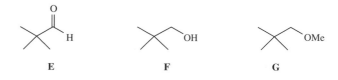

E **F** **G**

11. Natural Products and Synthetic Polymers

(1)

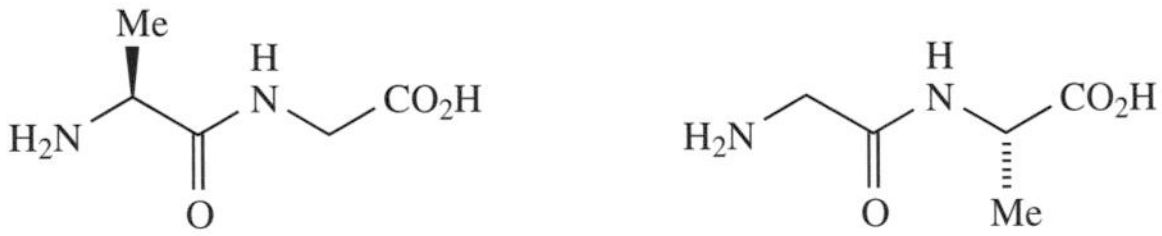

(2)   (a)   Disaccharide **A** contains a β-glycoside linkage.
      (b)   Two molecules of glucose are formed on hydrolysis.
      (c)   Glucose is an aldohexose.

(3)

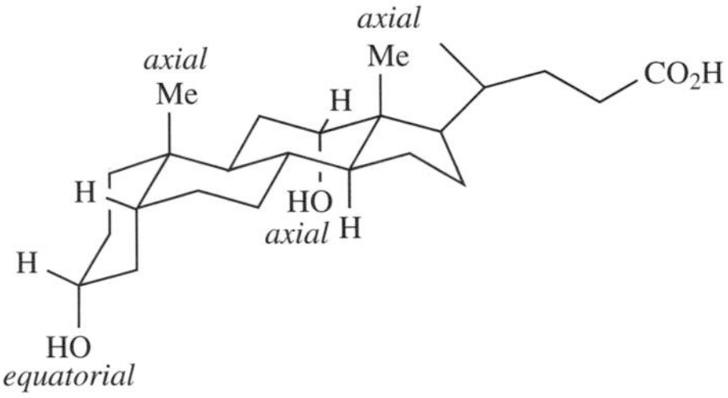

(4)   This is apparent when tautomeric forms of the amide groups are drawn.

C (uracil)

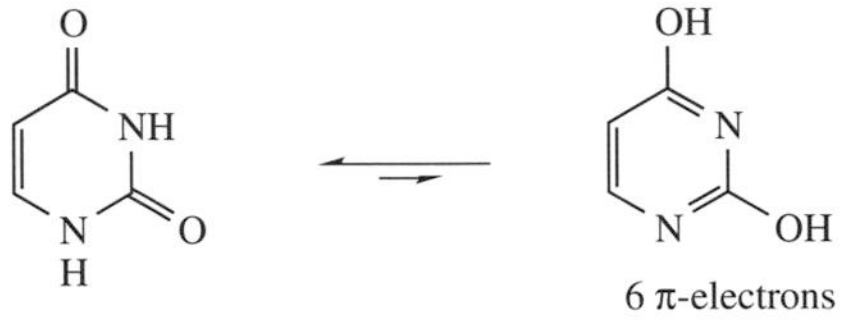

6 π-electrons

D (guanine)

10 π-electrons

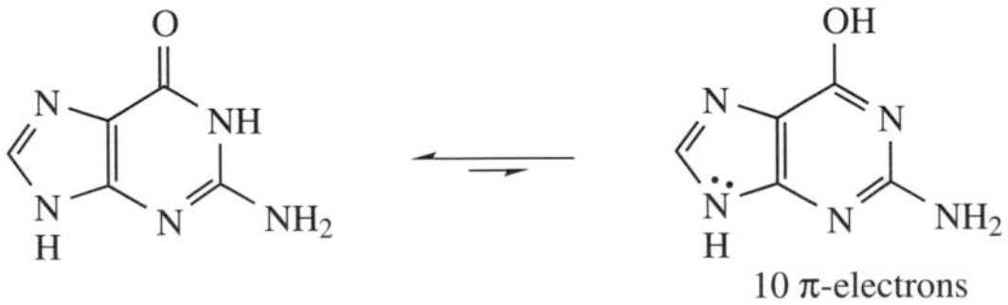

(5)   This is because radicals add to the less substituted end (or tail) of the monomer, and hence the head of vinyl chloride is attached to the tail of another.

head

vinyl chloride

head-to-tail

Cl

tail

Cl   Cl

poly(vinyl chloride) or PVC

(6) (a)

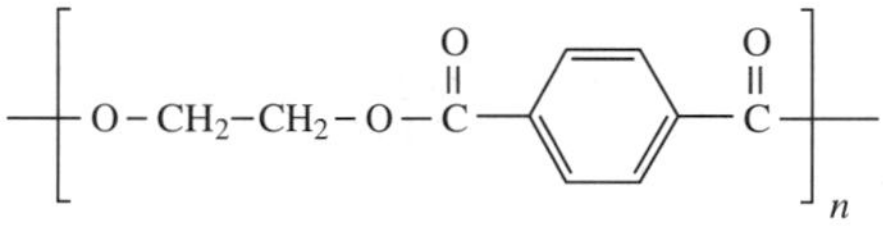

$$\left[O-CH_2-CH_2-O-\overset{\overset{\displaystyle O}{\|}}{C}-\left\langle \right\rangle-\overset{\overset{\displaystyle O}{\|}}{C} \right]_n$$

A polyester (Dacron)

(b)

$$\left[O-CH_2-CH_2-O-\overset{\overset{\displaystyle O}{\|}}{C}-\overset{H}{N}\underset{}{\overset{Me}{\left\langle \right\rangle}}\overset{H}{N}-\overset{\overset{\displaystyle O}{\|}}{C} \right]_n$$

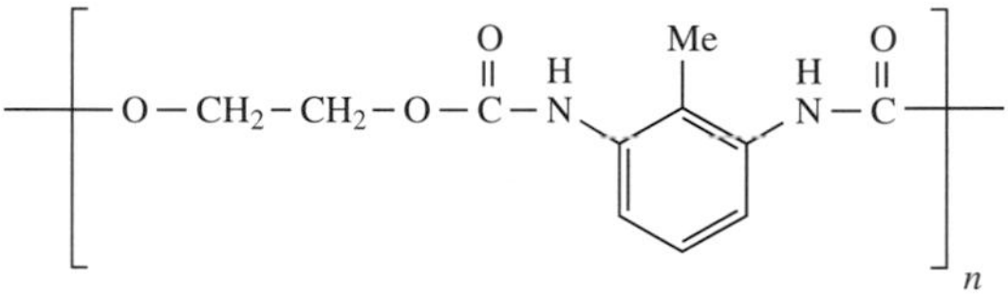

A polyurethane

FURTHER READING

There are a large number of excellent organic textbooks, which can be used to consolidate your understanding of the topics discussed in this book. Some of the more recent titles are highlighted below.

General chemistry textbook
C. E. Housecroft and E. C. Constable, Chemistry, 2nd Edition, Prentice Hall, Harlow, 2002.

General organic textbooks
P. Y. Bruice, Organic Chemistry, 3rd Edition, Prentice Hall, New Jersey, 2001.

J. Clayden, N. Greeves, S. Warren and P. Wothers, Organic Chemistry, Oxford University Press, New York, 2001.

H. Hart, L. E. Craine and D. J. Hart, Organic Chemistry – A Short Course, Houghton Mifflin Company, New York, 1999.

G. M. Loudon, Organic Chemistry, 4th Edition, Oxford University Press, New York, 2002.

J. McMurry, Organic Chemistry, 5th Edition, Brooks Cole, California, 1999.

Reaction mechanisms
H. Maskill, Mechanisms of Organic Reactions, Oxford University Press, New York, 1996.

M. G. Moloney, Reaction Mechanisms at a Glance, Blackwell Science, Oxford, 2000.

More specialist organic texts
See the Oxford Chemistry Primer Series, which includes many relevant titles including the following.

J. Jones, Core Carbonyl Chemistry, Oxford University Press, New York, 1997.

G. M. Hornby and J. M. Peach, Foundations of Organic Chemistry, Oxford University Press, New York, 1993.

M. Sainsbury, Aromatic Chemistry, Oxford University Press, New York, 1992.

Spectroscopy
S. B. Duckett and B. C. Gilbert, Foundations of Spectroscopy, Oxford University Press, New York, 2000.

L. M. Harwood and T. D. W. Claridge, Introduction to Organic Spectroscopy, Oxford University Press, New York, 1996.

APPENDIX 1   BOND DISSOCIATION ENERGIES

BDE = approximate bond dissociation energy in kJ mol^{-1}

Bond	H–H	H–F	H–Cl	H–Br	H–I	H–OH
BDE	435	565	431	364	297	498

Bond	H$_3$C–H	H$_3$C–Cl	H$_3$C–Br	H$_3$C–I	H$_3$C–OH
BDE	440	356	297	239	389

Bond	O–H	R$_3$C–H	N–H	C–C
Average BDE	464	415	391	346

APPENDIX 2 BOND LENGTHS

Bond	H_3C-F	H_3C-Cl	H_3C-Br	H_3C-I
Bond length (Å)	1.38	1.78	1.94	2.14

APPENDIX 3   APPROXIMATE pK_a-VALUES (RELATIVE TO WATER)

Functional group	Acid	Conjugate base	Typical pK_a
carboxylic acid	RCO_2H	RCO_2^-	4–5
phenol	C_6H_5OH	$C_6H_5O^-$	10
β-keto ester	$RCOCH_2CO_2R$	$RCOCH^-CO_2R$	11
water	H_2O	HO^-	15.74
amide	$RNHCOR$	RN^-COR	16
alcohol	ROH	RO^-	16–17
ketone	$RCOCHR_2$	$RCOC^-R_2$	19
ester	R_2CHCO_2R	$R_2C^-CO_2R$	25
alkyne	$RC{\equiv}CH$	$RC{\equiv}C^-$	25
benzylic alkyl	$ArCH_3$	$ArCH_2^-$	42
alkene	$R_2C{=}CHR$	$R_2C{=}C^-R$	44
alkane	R_3CH	R_3C^-	≥50

APPENDIX 4   USEFUL ABBREVIATIONS

Ac	acetyl [$CH_3C(O)-$]
aq.	aqueous
Ar	aryl
9-BBN	9-borabicyclo[3.3.1]nonane
Bn	benzyl ($PhCH_2-$)
Bu	butyl ($CH_3CH_2CH_2CH_2-$)
^{i}Bu	isobutyl or 2-methylpropyl [$(CH_3)_2CHCH_2-$]
^{s}Bu	sec-butyl or 1-methylpropyl [$CH_3CH_2CH(CH_3)-$]
^{t}Bu	tert-butyl or 1,1-dimethylethyl [$(CH_3)_3C-$]
Bz	benzoyl [$PhC(O)-$]
CI	chemical ionisation (in mass spectrometry)
d	doublet (in ^{1}H NMR spectroscopy)
DIBAL-H	diisobutylaluminium hydride ($^iBu_2AlH$)
DMF	dimethylformamide [$(CH_3)_2NCHO$]
DMSO	dimethyl sulfoxide [$CH_3S(O)CH_3$]
ee	enantiomeric excess (0% ee = racemisation)
EI	electron impact (in mass spectrometry)
Et	ethyl (CH_3CH_2-)
h	hours
IR	infrared
J	coupling constant/Hz (in ^{1}H NMR spectroscopy)
k	rate constant
mCPBA	meta-chloroperbenzoic acid ($3\text{-}ClC_6H_4CO_3H$)
Me	methyl
mp	melting point
MS	mass spectrometry
NBS	N-bromosuccinimide
NMR	nuclear magnetic resonance
Nu	nucleophile
m	multiplet (in ^{1}H NMR spectroscopy)
PCC	pyridinium chlorochromate ($C_5H_5NH^+ClCrO_3^-$)
Ph	phenyl (C_6H_5-)
Pr	propyl ($CH_3CH_2CH_2-$)
^{i}Pr	isopropyl or 1-methylethyl [$(CH_3CH(CH_3)-$]
Py	pyridine (C_5H_5N)
R	alkyl group (methyl, ethyl, propyl, etc.)
R,S	configuration about a stereogenic centre
s	singlet (in ^{1}H NMR spectroscopy)
THF	tetrahydrofuran
t	triplet (in ^{1}H NMR spectroscopy)
Ts	4-toluenesulfonyl ($4\text{-}CH_3C_6H_4SO_2-$)
UV	ultraviolet
X	halogen atom

APPENDIX 5   INFRARED ABSORPTIONS

Functional group	Typical stretching frequency (cm^{-1})
alcohol, O–H	3600–3200
amine, N–H	3500–3300
alkane, C–H	3000–2850
nitrile, C≡N	2250–2200
alkyne, C≡C	2200–2100
carbonyl, C=O	1820–1650
imine, C=N	1690–1640
alkene, C=C	1680–1620
ether, C–O	1250–1050
alkyl chloride, R–Cl	800–600
alkyl bromide, R–Br	750–500
alkyl iodide, R–I	~500

APPENDIX 6 APPROXIMATE ^{1}H NMR CHEMICAL SHIFTS

Methyl (CH$_3$) protons

Functional group	Typical chemical shift (δ) of methyl protons (ppm)
C–CH$_3$	0.9
O–C–CH$_3$	1.3
C=C–CH$_3$	1.6
R–O–C(=O)–CH$_3$	2.0
R–C(=O)–CH$_3$	2.2
Ar–CH$_3$	2.3
N–CH$_3$	2.3
R–C(=O)N–CH$_3$	2.9
R–O–CH$_3$	3.3
R–C(=O)O–CH$_3$	3.7

Methylene (CH$_2$) protons

Functional group	Typical chemical shift (δ) of methylene protons (ppm)
C–CH$_2$–C	1.4
O–C–CH$_2$–C	1.9
R–O–C(=O)–CH$_2$–C	2.2
C=C–CH$_2$–C	2.3
R–C(=O)–CH$_2$–C	2.4
N–CH$_2$–C	2.5
Ar–CH$_2$–C	2.7
R–O–CH$_2$–C	3.4
Br–CH$_2$–C	3.5
HO–CH$_2$–C	3.6
Cl–CH$_2$–C	3.6
R–C(=O)O–CH$_2$–C	4.1

Methine (CH) proton

Functional group	Typical chemical shift (δ) of methine protons (ppm)
C–CH–C	1.5
O–C–CH–C	2.0
R–O–C(=O)–CH–C	2.5
R–C(=O)–CH–C	2.7
N–CH–C	2.8
Ar–CH–C	3.0
R–O–CH–C	3.7
HO–CH–C	3.9
Br–CH–C	4.1
R–C(=O)O–CH–C	4.8

INDEX